TECHNOLOGY HOW DOMINATES HUMAN MIND

JOHN LOK

Contents

PREFACE

Introduction

Can technology influences human reading behavior changes to ebooks learning ? Nowadays, online and offline book shop competition is serious. Book readers have these both channel to choose to buy either electronic book or paper book to study. How can traditional offlince book shop achieve strategy to compete online book shop ? What are online book shop weaknesses or strengths? What are traditional offline book shop weaknesses or strengths? What is future book publishing development trend? How technology influence travel agent service changes to e-travel airline booking service from traditional walk in travel agent service?These questions will have suggestions to be given to book publishers to let them to learn more marketing strategies. Publishing industry competition is serious. Electronic books, newspapers will be popular to let readers have more reading method to choice. Also technology may excite future traditional new travel agent service changes to e-airline ticket sale method. How etravel agent service encourages traveler make air ticket purchase decision to go to any where to travel easily. How etravel agent encourges human buys airline ticket channel changes to online eticket buying channel?

Robot invention how influence human job behavioral change? Nowadays, robots may be applied to replace human some kinds of jobs, e.g. cookers, drivers, shopping center cleaners, even doctors, accountants or lawyers etc. different knids of occupations. Can robots dominiate human mind when robots can be developed to own themselves mind judgement abilites? When robots can learn human behavior that they can dominate human how to do any kinds of jobs, e.g. teaching teachers how to teach their students to learn new knowledge in schools. If robots can learn human any kinds of behaviors to do any kinds of jobs. Then, whether robots can dominate our societies to help us to change positive lives attitude to live more comfortable or change negative lives attitude to bring poor living environment, e.g. raising unemployment, reducing our salary. etc. dfferent negative social challenges. Whether (AI) robotic workers can be instead of traditional human workers in these different new markets to bring positive or negative impaction to change human job nature change? In recent years, machines had been used to be human's tasks in the performance of certain tasks related to intelligence , such as aspects of image recognition. Experts also forecast that rapid progress in the field of specialized artificial intelligence will continue. Then, it also brings this question: Does (AI) exceed that of human performance on more and more tasks to replace human jobs? If it is truth, will some of human jobs to be disappeared? (AI) will be instead of human some simple jobs, then unemployment rate to the low skillful and low educated workers will be increased.

Whether (AI) will be raised either production or performance or unemployment to bring human job market more advantages or more disadvantages? In my this book, I shall explain whether (AI) will bring benefits or disadvantages to human job market. I write this book aims to let readers make judgement whether technology can dominates human mind or human can dominates technology behavior in nowadays societies.

Prologue

Table of content

AI brings what jobs change

What does artificial intelligence(AI) mean?
What (AI) function is?
Can (AI) impact human job nature?

How can (AI) influence labor market?
How can human society job nature
to be changed to
artificial intelligent society?
Why does human need artificial intelligence machines?

How does artificial intelligence
influence future working changing

in automation employment and
productivity aspects?

Is artificial intelligence possible
to replace labor ?
Can (AI) technology replace human
labour nature of work?
Why can artificial intelligence satisfy
human needs?
Is artificial intelligence one good choice
for human future technological benefit? p.97-110

What is influence to (AI) job nature change to change
developed and countries countries economy
How can artificial intelligence technology influence economy?
Can (AI) influence gloabl economy
growth?
How can artificial intelligence
impact global economy growth?
How can (AI) influence GDP of high income
countries in the next ten years?
How can artificial intelligence
impact on workplace?
What is the relationship
between (AI) and (CRM)?
Can (AI) technology
impact on customer relationship
management (CRM) ? p.111-120
What is influence to (AI) job nature change to change
developed and countries countries digital economy

How can (AI) technology influence digital
economy?

What is the relationship between

(AI) and global digital economy development?

Could work activities in China be automated making in the nation with the world's largest automation potential?

How does (AI) technology influence the future of employment change?

How can artificial intelligence impact global economy growth?

Why will (AI) technology grow economic development ?

How can (AI) technology impact to global economic and social and psychological changes?

Will (AI) technology influence digital economy change to manufacturing industry ?

What is artificial intelligence potential benefits and ethical considerations?

How can (AI) technology influence to global health care economy development? p.121-139

Must Developed And Developing Countries Need Artificial Intelligent To Replace Human Job

- How AI help developing countries to communication and agriculture and learning and medical delivery development

- Emergency Response to developing countries' earthquake natural damage suddence occurrence predicting
- Smart AI Agriculture
- Medicine Delivery to developing countries' patients urgent need
- Assistance to reduce teaching work workload or psychological pressure to teachers in developing countries' schools
- Why does smart phone help developing countries

behavior which can influence the country's macro consumption desire?
Can technology influence human shopping behavioral change?

Why and how human behavior may influence the country's economic growth or recession?
Technology how impacts human behavior changing?
How and why employees behaviors may influence economy development?
Robots invention whether they can help organizations to raise efficiencies or inefficiencies? p.201-233

I

Technology how influence human reading behavioral change

shop different
development trend

Nowadays, online book publishing is one kind of popular sale method to global publishing. For example, Amazon publish is as a business model with many potential advantages, relative to a physical operation. It held out the potential of lower book inventing and distribution costs and reduced overhead. Consumers could find the books, they were looking for more easily and a variety book topic choices could be offered for sale. It can accept and fulfill orders from almost any domestic location with equal ease. And most purchasers made on its site would be exempt from sales tax. One Amazon strategy hand, it would have to make its returns and redress processes transparent and reliable, and offer other ways for clients to learn, as much about the book possible before buying. Future online book market development trend, such as Amazon,

Barnes & Noble etc. online book shops. How closely would their clietns find book ordering, as a substitute for visiting book stores? In fact, Amazon is global the largest ingle online booksellers and sells many other products. Otherwise, Barnes & Noble, have been market share diminsh obviously. In the future, Noble & Barnes both will have their market share diminish continue obviously. There are also many fewer specialty re lowest. Hence, it seems online and offline both publishing methods will be competitive. It brings this questions: What is the trend between online book sale channel, its size relative to offline book sales channel, growth rate and the charcteristcs of reders who by online in the future? How book market's online channels are economically different , due to e-commerce's effects on online book market and supply fundamentals? How an online book sales channel might be expected to change equilibrium market outcomes?

I believe online book channel based sale activity varies considerably on these aspects: Sales in manufacturing printing cost, online sale services and online demand print book sale book topic choices. Such as author online advertising, change more or less sale price, online paper book shippng cost, visa card discount or online book shop member card discount book purchase, what welfares to online book buyers are.

Why readers chooce to buy books from internet habitally? In tradition, online book buyers habitally hope to use the internet to buy. Generally, they have these characteristics: They hope to use the internet to buy electronic books at home, they enjoy to read electronic book from computer, it is in any regular capacity , not ncecessarily to visit book shops to find books to buy and they can search any electronic from internet, electronic book is convenient to read from computer or laptop when they catch transportation or going to anywhere. Usually, internet users are higher income, more educated and younger. It seems that education is a sizeable determinant of who is online, even controlling for income. However, gender does not seems to be a factor in explaining internet use. Moreover, many of book qualitative patterns are seen for online book purchases in general are observed for electronic book products on on demand printing book products in particular. Predition in future, many of the traditional online products , such as electronic or print on demand books, computer hardware , electronic airline tickets, saw more modest , but still substantial growth. In the future, online sellers trend to be newer online book stores and have less brand or reputation capital to signal or famous brand quality. These factors can create in online book sellers, which also often involve delay. However, there are many reasons for online book purchasing. The most obvious is that readers don't have opportunity where unobservably inferior point of electronic or demand on print book purchases.

● Pricing strategy in online and offline
book retailing

The book price represents consumer behavior on price. On one hand, the model contains two probability fuctions which render consumers' reservation prices for each individual channel. On the other hand, it is based on numerous book distribution which represent probabilities from and to each online or offline book store separate channel. Price strategy of book sale concerns how readers select a particualr book? Both offine and online book information seeking price strategies point out the challenges for information systems development. Hence, book price decision based on readers' age, e.g. children book price will be chealer than adult book price, due to children book content is usually simple and papers page is less. Otherwise, adult book content is more complicated or difficult to understand and page number is more than children book page number. However, online book store disadvantages are that : information system still often fail in supporting the users in causal leisure situations. In order to improve online book search system. Online book stores need to be better understood user strategies and performance and translate them into purposeful features.

A common analysis approach is to compare price and user strategies and interactions in the digital environment with those that occue in similar physical environment. If online bookstores hope to decide more reasonable electronic books or on demand printing books sale prices to compete with offline bookstores. Since, the physical environment (in this particular case bookstores) usually preceds the development of digital environments, processes and strategies from interaction in the physical environment have already stabilized and experiences can be translated into patterns for digital information system development. Thus, some only digital electronic bookstores , such as Amazon publish' disadvantages are : It lacks physical bookstore environment sale experiences. Otherwise, some owning themselves physical book and online book sale environment bookstores, bookstores that can compare only either paper books or electronic books bookstores to predict what the reasonable sale book sale price more easily.

Are these differencs between online/digital book discovery environments and offline (neighborhood bookstore) services? Are researching recommendation strategies differences between observable in online and offline book search sessions? In general, interactive users studies based on user interactions in a ISBS developed web-based book discovery information system are aggregated cross multiple researcher groups. In order to provide a realistic book discovery environment, book collection should be large and comparable to other book discovery systems ,such as online book sale. For example, Amazon library

book collection is used consisting of approximately 1.5 million books. Each book contains general metadata (title, authors, publisher, publication , year, etc.) subject metadata (classification, code), subject headings , user generated content (Amazon publish user reviewer, library thing user tags).

- How does India book market trend?

Thus, I believe that online or offline bookstore different book research method will also influence readers' preferable book choices, then their choices behavior will influence how many times to find the book easily. If the online or offline readers can find the book topic or author name or contents etc. information easily.

Then, the sale chance of the book will increase. Thus, price can increase more. For high population country, e.g. India, China . Does it have more sale chance, due to many people are living in these countries? What us online book store trend in India? Online book can let readers to buy new books and old books from internet, rent or borrow books from internet or access it in the form of e book, e.g. Amazon publish is the big player of online book business in India today. India where dynamic technologies like mobiles are prevalent, e-book readers may soon make into average household. Some of publishing houses which predicted that it would be long journey for e –books to become part of life needs to India readers. Thus, India will be one potential e book market. India is the third biggest market for English books. However, there are challenges of online bookstore in India. IN fact, online book market has changed the way reading consumer use internet for knowledge. Nowadays, people prefer e books are accessible anywhere, any time for creating flexible and secure online bookstore for online bookstores that need to concern to sell their e books to India markets because India readers shall concern visa card payment method where it is safe to pay to read any e books from internet.

Besides, online information searching has touched every field of human life. In the future, it is possible that purchased via mobile are clothing/footwear and e book or on demand print books. Also , due to e book is one kind of popular reading product to be enter India market. Currently, the online book market in India is offering exciting and renewed services to the internet users. India readers can accept to buy old books to read from online sale channel. Thus, India will be one new second hand online book store market to follow developed countries, such as US, UK etc.

- Trend and development on the global book market

Under the influence of internet, new media , social networks. The way in which search to satisfy our needs. Internet is the high technological search method to change at the level of products and services, such as e book (electronic book or demand on print electronic paper book) and online e book rent service , online library e book borrowing services. Thus, in the future, global book market will be popular on concentrating selling e books or online print on demand paper books more than general walk in offline book shop paper books sale only method. Due to, internet changes traditional readers' reading habits to enjoy to read e books from mobiles, laptops, desktops more than paper book reading. Thus, the global book market will be predicted online electronic book sale format more than visiting walk in book ship sale format. The digitalization of information enables us to bring into discussion today contents separated from the physical, materials, paper shapes of the book. Today, books could be found online, read online for free or downloaded as an e book in English or any other language. Practically, the book has changed from paper to electronic book. In until , the internet and the e book , the changes were extremely slow. Today, digitalization produces rapid changes to the entire system of printing, distribution and reading books. Hence, the global book market trend will be the major implication on publishes, distribution, authors and book consumers. The online competition brings major changes to traditional distributors, the bookstores, the author of independent distributors noticeable decreased. The number of big distributors' stores will decrease. For example, Amazon publish is the best known global selling books online. Although, it can sell e books and printing on demand paper books both from internet channel conveniently.

In conclusion, e book market will dominate global online electronic book sale market and the e book publisher number will increase. As the same time, the visiting walk in offline book shop number will decrease, due to readers have accept to use laptops, mobiles to read electronic books from internet channel more than reading paper books. It implies paper book publishers need to change sale method, e.g. adopting internet to sell print on demand paper books, or reducing paper book sale price to attract e book readers to choose to buy paper books to read.

- Web vs School campus book store development trend

Why do students choose to buy textbooks online? What factors motivate students choose online textbooks purchase? Nowadays, many online book retailers, such as Varsity books.com and Bigword.com ,. Amazon publish.com are now capturing more of the textbook online store market. What is motivating this behavior changes to student market , instead of children story market, entertainment or travel or sport book market etc. topic market. What causes students to choose purchase textbooks online ? Can likelihood to make purchases online by predicted by various social and personal characteristics of consumers? The online textbook purchase growth is allowing online retailers to capture a substantial portion of sales in some sectors. What motivates consumers to shop on the web? But, what if these factors are nor significant , such as better product availability, lower cost, as is that case when comparing on offline textbook purchasing. There is no significant price advantage to buy textbook online, it is there an availability issue, given that textbook can be purchased in the campus store (Foucault et al., 2000).

I shall assume that precious positive online .
purchase is positively correlated with the likelihood of an individual purchasing textbooks online. Hence, it influences why readers choose to buy textbooks online again. Following , other factor web consumers are likely shop online to save time and/or money, but what of those consumers who shop online when an equally time and cost efficient alternative is present. With regard to textbook purchasing, the time invested in researching for the appropriate books is likely to be similar, regardless of whether the student bookstore or through an online textbook. With regard to textbook purchasing, the time invested in researching from the time invested in searching appropriate books is likely to be similar: regardless of whether the student chooses to shop in the campus bookstore or through an online textbook retailer. If time from purchase until use counts, online textbook shopping could be considered less time

efficient than its offline counterpart. Due to the readers need to turn on computer to link to internet to read electronic books or wait the print on demand to buy paper books from the electronic book store web site to wait the paper books to post to the online book buyer's home. Otherwise, offline bookstores can reduce time spending to wait the books to be posted to the buyer's home, after who pay money to take the paper book from the bookstore immediately. So, the non-waiting post book issue is still the text bookstore's strength to attract students to buy.

Prediction of direction of electronic books future trend

What is future trend of electronic book publishing development? To answer this question, we need to know what benefits of (electronic books) can attribute to human's needs. Nowadays, electronic books (e-books) are one way to enhance the digital library with global 24 hours a day and 7 days a week access to authoritative information, and there enable users to quickly retrieve and access specific research materials easily, quickly and effectively. Evenm some ebooks publishers choose to let readers who can borrow ebooks to online readers to read from online libraries to earn profit. For example, Amazon publisher lets every Amazon readers only pay about US$5 per month. Then, who can borrow unlimited ebooks to read from Amazon publisher private online member library website convenently.

Thus, it is one ebooks online borrowing strategy to compette with offline book stores and public library and school library in publishing industry. Due to offline book stores lack borrowing books services to any walk in readers. However, some countries' publich libraries also have similar ebooks borrowing to read services. An an ebook providers' electonic online libraries, online computer library center has been involved in the selection, catalogue and distribution of ebooks. Library users can able to remotely search, locate and checkout ebooks from the library's online public access catalogues. Thus, ebook publisher will have another public library competitor which can provide similar ebook borrowing service to online ebook readers from public library websites.

It means ebook publishers need to adopt any attractive ebook library sale borrowing service strategy to attract public library readers. However, as with any new opportunity, new challenge utilizes the internet opportunity to deliver new book content is no exception, Integrating ebooks into the digital library has created challenges and opportunities for librarians, publishers and ebooks providers for librarians in this ebook library borrowing service market to earn extra ebook lending service income. Because, online borrowing service library can have ebooks borrowing service, then why online ebook readers need to choose independent ebook publisher individual borrowing book service website to replace traditional public library paper book borrowing service. The reasons possible include that the readers can borrow ebooks to study from ebook publisher individual library borrowing website at home conveniently, but it is possible that they can not find any paper books to borrow from public libraries which are the same ebooks to be borrowed from any one ebook store to read, also ebook publishers can let whose ebook borrowers to borrow unlimited ebooks to read and there are longer extend borrowing ebook return days more than public libraries borrowing book return days and ebook readers have no penalty when they return ebooks too late and they can choose to pay little borrowing ebook charge in the month, if who do not expect to borrow any ebooks in the month, who can choose to stop to pay borrowing ebook charge in the month. Hence, they can choose to continue to borrow unlimited ebook numbers from ebook publishers and they are permitted to return ebooks longer time to compare traditional public libraries. For example, when the ebook reader pay only US$5 ebook library service fee to the ebook publisher in the month , then who can borrow the number of ebook up to 50 maximum number in the month as well as who can return the all ebooks to the ebook library within 60 days, it is longer return days to compare traditional public libraries. If the ebook reader can not return all these ebooks after the return day of 60 day. They can permit to extend more 60 return days. After this another 60 return days, they only need to pay US$5 penalty to the ebook store. Thus, it is one attrative ebook library borrowing service strategy in this competitive book publishing industry.

There is no doubt that the same trends that adopts ebooks and e-readers to US ebook publishing market are having a similar effect in other countries as well, such as Mobile ebook or laptop ebook technical development of reading devices that provide an reading experience similar to that of reading an actual book, the increasing penetration of the internet in all areas of life, which is significantly changing reading patterns and reading behavior. The increasing extent to which ebook or demand on printing book consumers are open to new technological reading trends, for which in particular that availability of attractive mobile devices, such as smartphones, portable games consoles, and MPS players are responsible to ebook reader tools.

Future trend will be that publishers and authors need to build close digital cooperation relationship. Publishers, bookstores and device manufacturers should take the opportunity to provide the market now with innovative ebook publishing products. And authors should explore opportunities for digital distributions and support publishers in their efforts to publish content. Publishers should also design a giving strategy and attractive ebook sale website that attracts customers without undermining the value of content. A well-thought out pricing strategy may also help publishers and content gain new customers, those who would not have purchased a traditional book , but may be inclined to buy an ebook that costs less, offers additional features , and works on a digital device . They already own there, usually the ebook price compares to traditional paper book price which have similar content, ebook price will be cheaper them the similar content of traditional paper book sale price.

In the future, ebook publishers will need to position themselves as content providers, and not just the suppliers of physical books. They will have to make content available on multiples media, in multiple formats, on multiple platforms. This content may not be limited to the text of a book itself, it may also include videos and games. This additional content may lead to incremental revenue.

In fact, the only lesisure activities more popular than reading books were watching television, listening to music such the radio and reading newspapers and magazines. Thus, every one should need to choose to enjoy to do what kinds of leisure activities every day. For example, if one person chooses to spend much time to either watch television or listen the music and radio or read newspapers and magazines in the whole day. I believe that he will spend less time to read book in the day. Then, it implies that ebook or paper book readers , the paper book or ebook buyer number will be decrease, due to they spend less time to read or without any reading behavior in the day. Thus, how to persuade every one to feel that reading book habit is attractive or important which can be one factor to influence the paper or electronic book readers, even electronic or paper book buyer number. Thus issue will be an attractive topic to concern for every ebook or paper book publisher on book publishing industry. If these both kind of publishers can persuade any person to feel reading book habit can bring benefits to themselves. They will spend less time to leisure activities. Then, ebook or paper book sale number or ebook borrowing service income will raise in the future. Thus, these both kinds of publishers need to concern how to persuade people to choose to spend some time to read books habitually every day. Consequently, psychological factor will be one important direction to raise book buyer number in publishing industry.

What are the factors to influence sales and
marketing strategies for publishers?

I feel that how to predict book buyers which is driven by book buying experience and the publisher's credibility (loyalty) factors which will influence the any book buyer whose make final decision to buy the book from the publisher. As a publisher, a major goal is to extend whose readership and extend whose readers' influences, but where to start? How do publishers understand and serve diverse readers and decision makers in different countries? Whether can readers find the kind topic of book from publishers only, when find the kind topic of book from the university libraries or public libraries? Hence, due to offline and online publishing industry competition is high, global publishers will need to develop a sales plan to satisfy readers' reading taste. For publishers need to conduct book exhibition activities, visit different author's decision makers to research what who like to write negotiate terms to publish books with individual authors, secure sales and manage orders etc. different regulations of publishing to every author.

I recommend online or offline publisher ought concern how to publish every book before they decide to sel their every electronic book or paper book to any countries' readers. The marketing strategy includes to develop plan every book sale projection, SWOT (strengths, weaknesses, opportunities, or threats) to every book to be published to the country's readers to implement the plan. Book sales program, email communication marketing, lead generation to analyze the results, eg. every book purchasing trends, customer profiles, marketing sementation for every book to follow up and bedrief: Measuring ROI, setting priorities and develops tastics, finally customer needs analysis foe every book sale, it includes GAP analysis, ebook online library visits numbers to experience the ebook and focus groups. The, it is cycle to the develop plan again. Thus, if the publisher can have a better understanding of pricing strategy plan which can create price plan to be strengthed changes or cancelled for every paper book or electronic book sale marketing price strategies. Bringing potentially and disastrous reading experience to readers , this factor can be one good method to increase reader number and book sale price and sale number method. Then, the ebook or paper book publishers can make more accurate ebook or paper book sale price to every sale market, e.g.

US or UK which is better book sale market, which kind of book can be the popular to these either market, whether UK readers like to read ebooks more or US readers like to read ebooks more or US readers like to read paper books more or UK readers like to read paper books more. Thus, the ebook or paper book stores can gather these data to analyze whether what every book topic sale price is more accurate to achieve the highest sale number and income.

Consequently, more appealing offerings can be developed to broader every publisher's audience and enhanced whose every publisher's image, segments of reader research, e.g. reader age, book reading taste. This is a measure level of penetration of journals and identity opportunity for growth GAP analysis marketing strategies will be popular methods to future book publishing. Based on first hand, expensive visiting and surveying librarians around the world, examing factors unique to each country and culture and make to recommendations integrate in every publisher's communication plan. For example, ebook trends pecentage of ebok spending in online ebook borrowing libraries is a publishing extra income from ebook borrowing readers. It is such one part of the overall electronic book market share income in the electronic book publishing market. In conclusion, internet technological innovation can bring new publishing business chance to ebook development , but it also brings competition to traditional paper book stores. So, paper book stores need have good marketing strategies to win their new ebook competitors.

Reference

Foucault, B. Lery, N. Rifkin, A. & Silfies , 2000.
" Comparision of textbook prices by retailer and by college" working paper. Cornell University, Ithaca, Ney.

Factors Influence Reading Behavior
In Publishing Industry

Analysis of factors influencing online
newspaper reading behavior

Nowadays, online newspapers will be popular to let readers to read any newspapers' news from internet. It showed that for online newspapers reader's intention is influenced by performance expectancy, habit and the habit of reading a print newspapers. So, newspapers consumer personal reading behavior was influenced by intention and habit. Some reading behavioral researchers showed some reasons to explain why traditional paper newspaper readers will like to change habits to study online newspapers.

Hence, changing reading habit will be one factor to influence traditional paper newspaper reader individual reading behavior changes to online newspapers reading habit. In fact, high technological communication media will influence mobile phone and internet both new communication media causes. These new communication medias will bring new print electronic media causes, such as print newspapers, online book products. Some of traditional paper newspaper readers will choose to read any news from online newspapers. The reasons include free charge, reading at home in convenient, not need go out newspapers, online newspapers do not need the reader's hands to touch the black word paper newspaper to be dirty, and waste less time to buy every day to achieve economic benefit.

Every online newspaper reader will have this factor to influence whom to change traditional paper newspaper reading habit. The factor shows that attitude has a direct effect on intentions, and is influenced by performance expectancy and effort expectancy or related personal online reading acceptance conceptions. Because of whose acceptance of online newspaper reading attitude is as an important in technology user online newspaper reading attitude was included.

Additional, every online newspaper reader self-efficacy and anxiety are expected to be minor issue to influence the online newspaper reader to change whose attitude to choose not to internet tool to read of an online newspaper. However, different age reader either he/she is young or old age factor will have influence whom to choose online newspaper to read, e.g. old age readers will feel difficult to apply internet technology to read newspaper, otherwise, young age readers will feel easy to apply internet technology to read newspaper. So, the old or young age traditional paper newspaper readers, when the acceptance new technological online newspaper to them, they will adopt online newspaper reading attitude to replace traditional paper newspaper reading habit more easy. So, their acceptance new technological of online newspaper reading attitude will have a direct effect on online newspaper reading intention and are influenced by both paper and online newspapers reading enjoyment performance expectation and online newspapers reading effort expectation, when their expectations were needed to be satisfied more these past traditional paper newspaper reading experience. Moreover, past paper newspapers reading behavior and habit should be noted. Then, these two expectation factors will encourage or persuade the traditional paper newspaper readers change whose reading attitude, reading habit and reading behavior to read online newspapers. Hence, the online newspaper readers' psychological factor will influence whose traditional paper newspapers readers whose reading behavioral changes. Also, it means that expectation factor will influence the traditional newspaper readers to change whose counter intentional paper newspaper reading habit.

However, online newspaper will bring much knowledge to compare traditional paper newspapers , e.g. real newspapers news data, more meaningfulness news, providing the nature of visiting a news website, which can let online news readers can feel different read model primary on frequency with relatively little spread in the amounts of time spent at the site.

What are the main factors to influence online newspaper reading behaviors? Same testing indicates for moderation by the online newspaper age, gender and online newspaper reading experience will bring the online reading newspapers habit influence. The testing also indicates male gender and young age group , this group likes to apply internet to find or seek or search any news matters. Hence, this internet user group will bring to have interest to read newspapers from internet channel. The reason is possible because this young male internet users like to contact new technology, e.g. internet. They think the online newspaper is useful and it is more useful to read the online newspaper to compare to paper newspaper.

The two reasons : liking to contact new technology and feeling the online newspaper is more useful which can support why young male online internet users feel to expect reading online newspaper expectancy were more concrete.

Additional online newspapers usefulness are more considered on unclear concept to explain why this reader group feels more like to study online newspapers. What exactly is the usefulness of reading an online newspaper?

The testing also indicated that some online newspaper readers responded to use the online newspaper to feel natural, it showed to be related to attitude as well as to habit , which seems to hold face validity as a natural feel can be considered on attitude on the online newspaper. So, online reading attitude and online reading habit can reflect why man young male like to read online newspapers more than paper newspaper reason.

Another reason indicated that when the young male readers want to read the news, the online newspaper is an obvious choice for him/her. So, many online newspaper young male readers had felt online newspaper is one another newspaper reading choice to replace traditional paper newspapers.

In conclusion , free charge online newspaper is not the main factor to influence both traditional paper newspaper readers to change their reading habit to choose online newspapers to read suddenly. There are other factors to cause them to choose online newspapers to read, such as more usefulness feeling, contacting new technology, online reading habit, positive online reading attitude etc. different psychological factors which will have more influences to cause their paper newspaper reading habits to be changed. Hence, the free price economic gain actor must not only one main factor to persuade readers to choose online newspapers to read.

How electronic versus traditional print textbook influence of university students' learning behavior

When one university student was accepted by electronic text book learning channel to replace traditional paper text book learning channel (methods). Electronic text book will bring what positive or/and negative influence to impact whose learning behavior changes. For example, electronic text book learning method will bring positive impact to raise the student's examination grades and perceived learning scores or it will bring negative impact to fall down the student's examination grades and perceived learning scores. The mean scores indicated that students who choose to text books for their learning aim. It will have significantly higher perceived affective learning performance and examination results. Thus, the purpose of student learning and teacher teaching method, every university needs to examine whether it is efficient to raise student learning effort to replace paper text book learning method in any learning environment, e.g. many students and one teacher classroom learning environment or the independent student learns himself/herself at home learning environment or library learning environment.

Can text book reading tool bring absolute advantages to university students or bring some disadvantages to them? When a student needs to apply e-text book to learn, who needs access e-text book in a static location, such as a computer or on a mobile device. So, the e-text book in a static location factor, it will have influence to each student reading or learning behavior to bring negative and/ or positive both impacts.

The e-text book was distributed on a CD and installed on a located computer. This limited the user to accessing the e-textbook in a single location and eliminated the potential access to the e-text book on due to the lack of mobility. So, it seems that the location of limited to e-text book will bring negative impact to let the student can only learn in a fixed location because he/she will feel difficult to move heavy computer to other places to learn more than on paper text book. So, e-text book location can not allow the student to leave the classroom to learn more easier if he/she had chose to use to computer to install the CD to learn in the classroom. Supposing the student 's teacher needs the student often to leave the classroom to discuss any matter suddenly, it is not very convenient to the student to use e-text book to learn because he/she can not move the computer to leave the classroom with him/her easily. Then, it will be possible to influence the student can bot be attention to read the e-text book, when the teacher needs the student to leave the classroom (none book bringing) to discuss any time any time immediately. Otherwise, if the student used one paper book to read/ learn in the classroom, if the teacher needs whom to leave the classroom often to discuss immediately. He/she will feel convenient to learn because he/she can bring the light paper book to leave the classroom to discuss with the teacher in any location easily.

Hence, it seems e-text book learning will bring not convenient fixed location learning environment to every e-texting learning student in classroom, when, he/she needs often to leave the classroom to discuss with the teacher any time.

Other disadvantage of e-text learning will bring students feel difficult in possible. In the past, some learning researcher experiments indicated results demonstrated that student participants in both groups had similar recall and ability to reinterpret information suggesting that retrieval of information is not effected by kindle e-book reader e-text book, a tabled computer e-text book or a print version.

Hence, it seems that e-text book can not help or assist recall the student's learning memory to remember the e-text book content more easier. Due to it is one e-text book machine, the student will fell to difficult to find any unclear or important information in any page(s) to write for learning/reading record more easier than one paper text book.

Another disadvantage of e-text book is the inefficacy or inefficient reading challenge to the e-text book reader. The efficacy of e-text books in a higher education environment will be one interesting discussing question. Passage length is one difference that impact the results. Studies involving shorter reading sessions indicated no substantial variance with respect to reading comprehension and understanding.

Conversely, studies involving longer reading passages indicated prior comprehension, when reading longer e-text , eye fatigue and mental workload are also concerns. Hence, e-text book reading will be possible to let students feel eye fatigue and mental workload in their reading process.

Due to machine e-text book words are more small size and unclear more than paper text book words to print to let students to read every words or sentences in computer. Consequently, studies indicated that e-text book readers need to spend much nervous and time to read longer and poor comprehension in whole e-text book reading process. When, university students need to spend time to read longer e-texts from computers. For example, they need to choose to reads hundreds of papers of e-text books on a screen, whether on a computer or handheld electronic device compared to print versions may contribute to eye fatigue. The consequence, eyestrain and mental fatigue could be poorer comprehension and have a poor eye, nervous health influence and every student's e-text learning behavior can bring negative reading habit to whom, when every one need to apply desktop or laptop or mobile electronic tools to read any words from e-textbooks. Hence, it seems e-text book reading method has possible to bring poor health challenge to every student.

So, above all these e-learning reading factors to bring this question: Does e-learning influence the student's negative reading behavior to cause poor final examination grades results? In fact, every student needs to change whose traditional learning method from paper text book reading habit or reading behavior to e-text books. He/she needs to change whose reading habit. He/she must need to spend long time to accept how to adopt this kind of new technological reading method as well as effect may change through a new technological learning experience itself and impacts the acquisition of knowledge leading to reading behavioral change.

As I indicated the e-learning will bring poor health and poor nervous negative influences when the student often needs to apply electronic product to read long time. So, it will be possible to influence the student health to be poor to bring examination low grades results in possible be cause he/she has poor health to exam.

It is possible that it has relationship to influence the student to exam low grade between e-learning habit and poor health causes. The reason is based on that efficacy of textbook format is defined grades. I assume that all these negative e-text book reading factors can influence every e-text book reader's health to be poor when he/she needs often read e-text book s to cause long time reading habit. So, I mean that e-text book reader individual health changes to poor, it is only depend on how long time e-text book reading habit factor. So, if he/she only spend less time to read e-text books and he/she also has habit to read paper books sometimes. Then, he/she won't be influenced to be poor from e-learning method easily.

It means that it has none direct relationship between less time e-text book reading habit and low examination grades result. Hence, low examination grades result to the student, it only depends on long time e-text book reading habit and the student's long time e-text book reading habit needs to confirm that whose long time e-text book reading habit causes poor health to the student effect. Why do I believe that efficacy of textbook format can influence the student's examination grade? Based on above analysis, the e-text book reading format and paper book reading format is very different. For example, the efficacy of textbook is very different between paper text book reading format and electronic text book reading format. Such as one sickness student needs to spend more nervous to read one e-text book more than one paper text book . This reason is because machine reading method is difficult to compare paper reading method. When the student has sickness, who must need to spend more time and nervous to read one e-text book more than one paper text book. If my assumption is right, then the e-text book sickness reader's reading efficacy to each paper must be poor to compare the paper text book sickness reader , due to the sickness e-text book reader needs to spend long time and much nervous to read each paper more than he/she chooses to read one paper book. When he/she is sickness to finish whose reading . Due to his/her memory will be poor and tries, when he/she feels sick, so whose reading effort must be poor when he/she needs to apply computer tools or mobiles to read.

How to change future e-reader
study habit to feel better

Nowadays, publishers, internet bookstores manufacturer e-readers have high expectations for digital future of book industry. If they expected e-book publishing industry success, they need to considerate how to assist to future e-readers to let them to feel whose reading habit to be better in order to persuade or attract them to choose to read e-books more easily, due to doctors indicated that long time e-books reading will cause eye poor health and tired and poor nervous reason in possible and paper book price competition and more topic choice reason. It is one value consideration question that e-book publishers need to considerate.

For example, in the US Amazon publish has improved the reading market by producing e Reader that is easy to use and making it easy for clients to purchase a wide variety of books at competitive prices. It will bring digital reader technology as an opportunity to open new target markets and create new e-readers. The question is how Amazon publish , such as e-book publishers change future e-reader reading habit to feel to choose e-books reading method is better than paper books reading method. The successful factors may include as below:

E-book reading market is similar to e-music listening market. They need every e-book reader and/or digital music listen listener to discover why to apply this kind of new digital technology reading or listening method which is better to enjoy to read every e-book content and/or listen every digital music song in order to adopt new listening and/or reading digital technological learning habits or experiences. So, this new digital technological reading or/and digital music listening process, every e-book reader or digital music listener needs to learn how to adopt this kind new digital reading or/and listening products to change from his/her traditional paper book reading or/and CD/DVD music song listening method to this new technological digital reading or listening methods from computer tool channel.

In this changing habit process, every e-book reader or/and e-music listener needs to spend some time to learn how to apply internet technological tool to help whose to read digital book or listen digital music from computer channel. So, he/she must attempt to change whose habit from traditional paper book reading habit and/or CD/DVD listening music habit to e-book reading habit and/or e-music listening habit.

Furthermore, e-book publishers also need to know whether which kind of book topics are be favorable popular to be chose to read for either student reader target to read or mature age reader target to read or old age reader target to read. Who will purchase the topic to read to be e-Reader? Will they be designed to appeal to be a group of e-reader customers or only to those who have a high degree of comfort with technology to enjoy e-reading method? Will people who read once in a time be purchased by the small group of e-reading clients who buy and read a high volume of e-books? What reasons, readers will choose to read the topics of e-books more than paper books? Will publishers be able to use e-books and e-readers to extend the many different age e-reading clients, e.g. young, mature, retirement, old, student age e-readers? Will publishers ever more to all readers are only choose digital e-reading model habit or who are a half digital e-reader and a half paper book reading habit clients to them?

Hence, one successful digital publisher needs to consider how to persuade every traditional paper book habit readers to change their reading habit to read digital e-books from computer. Because changing habit is one challenge to influence the e-book publisher 's e-book reader number. How to persuade the paper book reading habit readers to change whose attitude to choose to read e-books , it is one considerate question to every digital publisher? Some readers may feel difficult that who needs to learn new knowledge how to read e-books from computer tool, e.g. old age reader group. This reason will influence they still choose paper books to read in habit. So, any e-book publisher has responsibility to teach new digital technological knowledge learning method to let the e-book desire readers can feel easy to apply internet to read e-books from computer tool.

Another factor is e-book price, normally every e-book price will need to be sold cheaper to compare the similar paper book topic in order to persuade paper book readers choose to buy the similar topic of e-books to read more easily. Because if the reader discover the e-book topic is similar to the paper book topic contents, but the e-book price is charged high than the similar paper book topic content, then he/she will possible to choose to buy the similar paper book topic to read.

Another factor concerns how to raise e-books attraction. E-publishers will need to position themselves as content providers, and not just to be similar to the suppliers of physical books. They will have to make content available on multiple media, in multiple formats, on multiple platforms. This content may not be limited to the text of a digital book itself, it may also include audio, video, image and sound speaking digital books to attract e-readers' attention.

Another factor is that I recommend e-book publishers need to let all e-book readers to feel reading e-books are leisure time habit to let them to enjoy life every day in popular. Intention is such as good tool for anyone to apply to entertainment, for example people linked using internet to read books, watch movies, play video games from computer tool. They are some main points. They have same main points. They tend to spend leisure time with electronic media, such as apply internet to watch television which is such as to apply internet one more choice to assist readers to read e-books from computer media tool conveniently at home.

However, this is one example e-book attraction point to e-reader. Every e-book needs have e-pub files to allow readers to control the size of the text or their computer screens. If the e-reader feels the text is small size and computer screen is small size in difficult to read. Then, he/she can use mouse tool to change the e-book text number to be high number, e.g. from 18 to 20 or more number and he/she can apply mouse tool to move the computer screen to be wider more easily. Hence, it is e-book attraction point to e-book readers to feel when he/she feel the text is small size to read in difficult. Otherwise, every paper book print text(word) size is fixed, all word size can not be changed to read and every paper book wide size is also fixed. All it is every paper's unattraction point to every paper book reader.

Consequently, every e-book publisher needs have its attraction point to let its every e-book reader feels it is different to the other paper book publishers. It needs to solve these challenges to let its every reader to accept to choose its e-book reading channel. The challenges may include how to let the e-reader feels its e-book reading media can provide a more comfortable e-reading experience to compare other e-book publishers' reading media, how to let its e-readers feels its all e-books can provide one precise and stable e-book reading characteristics, how to let its ebook readers to feel its every ebook displays does not require any background lighting and one easy to read, even in direct sunlight environment, and it e-reading tool can spend less energy from laptop battery or desktop electricity to compare other e-book publisher reading tool, it means that the e-book publisher's ebook reading tool can provide a recharged power desire which can be used for several thousand pages or seveal weeks e-reading function. Hence, it the e-book publisher's e-book reading tool can provide more clear words and text image as well as less electricity consumption function to let every e-reader to read to compare other ebook publishers from laptop, destop or mobile media. Then , the ebook publisher's competitive effort will raise to win its other ebook publishing competitors. However, any ebook publisher needs have attraction points to persuade its ebook readers to read its any ebooks feel comfortable and providing fun ebooks choices and easy to read its every ebook text more clear if it expects to win its competitors in ebook publishing industry.

Factors influence child reading habit

Reading failure is a serious educational problem to influence every publisher success because if the child chooses to buy its books to read, but its child readers can not feel its books can help them to assist their learning success or failure examination or low grades result. Then, it will influence its child reader number to be reduced. However, the factors cause reading failure, it is not only considered to the publisher's poor book content quality factor, it can include the other factors such as: It is simply be attributed by poverty, immigration or the learning of English as a second language. What factors will influence child read in wrong habit to bring reading failure, even learning failure in effect? It is one question to every publisher needs to know in order to avoid they feel failure examination emotion after read their e-text books. Hence, how to design every text book content is one important issue to ever publisher.

A study by Yankelovich found most children are reading, but they are not reading enough. It indicated only about 3 in 10 children can be classified as high frequency readers who read books for fun ever day. Age 8 children are less to see benefits t oreading for fun, girls are more likely boys to have positive attitude about reading and feel fun. The benefits of reading are evidenced by the attitude of high frequency readers to achieve future learng success. More than 40% of children ages 5 to 8 say they are high frequency readers, by ages 9 to 11 that proportation drops to 29%. Almost half of the 15 to 17 year old (46%) are low frequency readers compared with 14% of 5 to 8 year old age. So, this study investigation reflected that building good learning habit has relationship between frequency reading and feeling fun to read to every child. It seems that one fun content book can attract the child to read the whole book all content really. So, publisher needs to consider how to design and write attractive content books to let every child to read.

What factors cause every child feel barriers to read? Some investigations indicate that young children tend to maintain high expectations for success, even in the face of regarded failure, when old students don't, also to older students feel failure following high effort appears to carry more negative inplications. Moreover, all students individual attitude about their capabilities and their interpretation of success and failure is further factor to affect their willingness to feel fun to read in themselves learning proceses.

So, it concludes this fun book content design method can persuade young people feel fun to read really. Moviated readers hold positive benefits about themselves attitude or reading habit which will bring positive and attractive reading emotion to influence them.

What are the book publishers and teachers' responsibilities to improve student individual negative habit to have positive reading habit or positive learning attitude? The ultimate goal in teaching and reading book is to raise students comprehend te ideas in a piece of text as they need. So, any publisher has responsibility to publish one fun and meaning book in prior, because every teacher will teach whose students by the text book content. If the text book content is fun and attractive and meaning, then the teacher can teach to let every students to learn more easily.

Training every student owns good reading habit which can help whom expand their thinking skills, learn to concentrate and enlarge their vocabulary and effectively better their learning environment. The good reading habit ought be trained from the child stage in beginning. So, when the child has is growing up, when he/she is needed to go to primary, secondary, even university to study, he/she had been built good reading habit from the publishers' fun and meaning book content influence in order to let they further learn any new knowledge to feel more easily. So, publishers have responsibilities to sell fun and meaning content books to let every child to read to raise whom reading interest to further young and mature learning stages.

However, the problems, children experience learning to read are often not related to their ability to learn, but to their awareness. Their ability to hear the English language and their expose to the English words. So, repeating to spell the English words will assist the child to raise memory to remember to write the English words more easily. So, book publishers have responsibilities to express every book content to attact child readers to feel interest or fun to learn to remember to spell every word as well as teachers have responsibilities to train students how to hear the words, he/she assist every child to learn to spell the English word more easily. So, teachers ought often speak every word or speak every sentence loudly from every book content to let students to listen easily in order to let they can raise every word memory more easily.

Consequently, instead of child's parents and child himself/herslf has responsibility to help the child self to build good reading habit, teachers and book publishers have also responsibilities to help them to build good reading habit because fun and meaningful books which bring more attraction to influence every child to read, when the book is fun and meaningful , then the teacher can follow its content to teach whose students to attract them to learn more easily. However, the good reading habit includes elements of reading comprehension to every book content , such as: identifying and summarizing the main idea, comparing and contrasting, identifying supporting facts and details, making influences and drawing conclusions, predicting outcomes, recognizing fact and opinion, identigy cause and effect recognizing sequence of events, identifying story / case elemetnts, such as main characters, settings, conflict, and resolution, identifying the another's purpiose and point of view, interpreting literary devices, such as imagery , symbolisms.

On conclusion,I believe that technology may influence human changes reading behavior from paper books to ebooks reading behavioral change. Moreover, technology ebook may influence reader psychological mind change to like to choose ebook online reading behavior more than traditional paper books reading behavior. Hence, technology may influnce human reading behavior and reading mind change from traditional paper books reading method.

II

Technology how influence travel agent changes to e-travel agent service

The main factors that can affect their strategies to reduce airline costs. The main factors include route structure, type and characteristics of the aircracft, cost of labor and management quality, which will influence whether which airline routes are the most suitable to let online travel agents or offline travel agents to help them to sell paper air tickets or electronic air tickets to attract travel consumption more easily.

Thus, a cost-related strategy is the main important factors to influence travel consumption choice between online or offline travel agents. For example, considering that advantages in costs is an important strategy for carriers to remain in travel transportation market.

The deregulation process of travel markets and increasing opportunities for competition have created excess capacity in many markets that causes lower rates, even with its rising costs. Thus, the travel strategic costs management as well as travel consumers that their behavior under different influences can bring competitive advantages over travel players.

Cost reduction in the travel market -based industry is a very important way of being competitive between offline and online travel agents, when facing travel air ticket prices decreasing for every trip. So reduce to total travel cost, e.g. fuel, maintenance, labor etc. is relevant, but the influence of ech component on every total trip cost depends on factors that are related or not to airline opertion. For example, some airline can adopt the lowest cost model to sell air tickets from offline or online travel agents which compete for travel passengers with traditional modes as self driving road transport trip in large areas

Nowadays,many online or offline travel agents have interest to find what of countries domestic travel market, such as US, UK domestic travel market.

However, the decision about the relevance of one cost is not a simple matter. The effectiveness of reduction of each item that comprises the total cost of airline can change over time, depending on both the business model and the scope of the airline company or online /offline travel agent company as well as external factors.

However, there are three types of competition advantage between online and offline travel market: They are such as agility, differentiation cost and the differentiation may be related to a product of superior quality, higher value f the brand or the company's positive reputation. Such as the online travel agent's providing the different airline cheap air ticket price and kind of trips to provide to travel consumer consumer comparison or the offline travel agent's famous brand or positive reputation to let travel consumers feel travel agents can provide many actual trip package to let them to compare by oral clearly. Thus, the online travel agent's weakness is lack of travel agent individual exploration to let every travel consumer to understand every trip package more clearly.

But online travel agent's strength is it can provdide one website

to let travel consumer attempt to compare different trip air ticket and/or hotel price to make personal travel pre-booking decision at home. The another advantage is related to techniques that reduce production cost, making it is possible to offer cheaper air ticket, or hotel room rents, or cheap trip package, than the competition. Such as online travel agent can sell more cheape electronic air ticket price to compare traditional offline travel agent's paper air ticket price.
Finally, agility refers to the speed which the company responds to market demands. For example, if the online travel agent can make statistics to analyze how many online travel consumers to choose to buy which airlines' electronic or paper air tickets, e.g. which airline trip destinations and trips and hotels choices are the most popular attraction to them. Then, the online airline has possible to respond to provide to the most popular airline trips choices, electronic air ticket price comparison choices and hotel rooms prices choices to attract many online travel consumers to enter their online travel websites to choose different airline electronic tickets to buy or pre-book hotel rooms from travel agent websits. Also, if the traditional offline travel agents cn attempt to gather every travel consumer's destination trips, hotels , airline paper or electronic ticket prices enquires to make statistics to make which travel trip journeys or destinations and airline paper travel ticket prices are the most popular. Then, it is possible that they can respond to every travel consumer individual demand more to attract whose travel agent choice more easily.

1.2 Airline travel agency AirAsia in the domestic airline low cost strategy

There are three major characteristics of the airline industry namely is product nature, its expenditure structure and its market entry conditions. Airline agent's product is homogeneous or undifferentiated , causing significant competition in airline domestic travel or foreign travel both markets, which are free from regulations and economic barriers. However, high capital and operating expenditure is another important characteristic of the airline industry. Aircrafts, airlines' major capital expenditure are very costly to acquire . For operating expenditures, aviation fuel and labor make up the two major costs in the industry.

Another important characteristic of the airline industry is the conditions for market entry, which differs between international and domestic airline markets . In the international travel market, airline travel agency entry is very difficult as international flights and routes are the results of regotiations between governments . On the other hand, in the domestic and regional travel market, travel agency entry depends on the level of deregulation or liberalisation. More and more countries, however are opening up their domestic travel markets for more competition. In addition, government plays an important role to regulate the travel markets and existing players may significant influence over now travel agent entrants.

In fact, the mjor factors influence to international or domestic travel consumption increasing numbers are the global economy and safety issues, instead of other different economic factors, such as travel destination choice, electronic air ticket or paper air ticket price, hotel price , the country's political change, e.g. war occurrence, bad weather , e.g. very cold or very hot etc. different factors infuence. Because generally , the world or any region of it is in an economic crisis or depression , the demand for airline services will fall. The late 1990 year Asian financial crisis for example, resulted in minimal increase in the number of worldwide airline passengers incrased only minimally from 1997 to 1998 year. Another factor of influencing the travel passenger number to be decreased, it concerns safety issues are also an important driver of the travel industry, which is subject to very safety standards to influence travel passengers' travel choice to the country. In addition, they are also unexpected safety related events, such as the 11 Sept. 2001 year tragedy in the US, which caused reduction in passengers . The increasing popularity of low cost airlines is the newest trend in the airline industry if which hope many passengers choose to buy whose electronic air ticket or paper air ticket to catch which planes to fly from online travel agent or offline travel agent channels.

The rise of low cost airlines, such as AmericaWest, JetBlue and Airtran in US, Ryanair and EasyJet in Europe and Vigin Blue in Australia. The share of low cost airline strategy is popular in the US and European airline market. For example, the Southwest airline low cost strategy is the basis of most low cost airlines operations. The key of the strategy is to reduce costs when at the same time offering low prices to passengers. History showed that the low cost airline strategy is easy to replicate , but difficult to implement successfully.

However, I suggest airlines need to know what functions which can attract passengers to chose to catch their planes to fly if they expect to rise passenger numbers. For example, the critical function of the Malaysia airline travel is to connect the major towns and remote interior areas within East Malaysia, which has poor road systems and limited availability of other significant means of transportation . In constrast, West Malaysia has more developed and extensive rod and railway systems.

Therefore, airline travel is not the main mode of long distance transportation. It implies Malaysis airline ought concentrate on focusing short distance transportation strategy for passenger benefitical choice function. For example, a new small Malaysia airline serving one or two routes may enter easily. Otherwise, a larger airline servicing multiple routes may be harder to enter Malaysis airline market. It also means access to capital and labor are the major obstacles for new airline entrants to Malaysia airline market. Thus, small airlines into a larger airline is probably more likely to be successful as in Air Asia's case to Malaysia airline market.

Thus, the airline low cost strategy competition positions include very low or minimal pressive from other airline similar service substitute products, low or medium power of airline similar input suppliers. In conclusion, low cost airline strategy is a god method to be attempted to win competitors in airline market.

1.3 How consumers select travel service between online and offline mode in travel industry

Nowadays, the travel industry is operating through two different modes, online and offline respectively. It involves the identification of the competitive strategies adopted by the tour operators. For example, it was found that e-retil travel is platform that is bringing two market forced the demand and supply tour operators and the customers together, and both parties and more inclined towards online mode in near future. Tour operators are gaining by operating at low cost and increasing their business reach when customers get what they desire as per their convenience. For example, many tour operators had promoted tourism destination through website that allow user to use interface for booking transporttion, foreign exchange etc. However, the role of travel operators (agents) should be assisted any airlines to promote their travel package service by internet more easily , such as tourism destination , arrangement of hospitality, restaurants, transportation tools during their trips.

The reasons why consumers choose online travel service include:

Firstly, it is online researching hospitality service. Online travel websites can provide many different accommodation furnitures, such as seeking hotel locations, rooms prices comparison, prepaid hotel rooms by visa card payment transaction method, range from luxury five stars deluxe ctegory hotels to small guest houses. The primary need of tourist is to find a place for residing in foreign country or domestic country to ensure whose safety and relaxing needs. Online travel website channel can help whom to find a place , according to his/her needs and paying capacity in the most shorten times.

Secondly, it is online restaurant (food and beverages researching) service. Full service restaurants are divided into two categories, fine dining and casual dining restaurants . Fine dining restaurants are usually located in the premises of luxury hotels, provide high quality food at premium price with good ambience and highly trained professionals. Thus, travel consumers can also compare the different restaurant food price and seek where is the restaurant and find.

What food taste of food supply from the travel agency or travel operator website easily 250 + tour operators are registered with the ministry of tourism (website of tourism ministry) , and the major players in the industry are dealing online and are dominating the travel industry. The major online travel players are Thomas cook, Cox and Kings, make any trips, clear trip, gatra.com and Expedia.

The tour operators whether online or offline offers a large number of services to the touists including customized package where the customer selects each element of the tour package, speciallized tourism package and complete tour guide package.

Nowadays, the tour operational travel (agents) are working through two different modes: offline online . Big brands with luge investment are dealing online and enjoying low cost benefits and huge profit margins. When the small tour operators have their market niche and managing have their market niche and managing their profits by dealing offline.

It is generally prefer offline mode that is the opportunity for small capital investment or employee number for tour operators. But the large scenario is changing as with the usage of internet by the tour operations have given convenience to the customers and now the customers of modern age have started developing preference for online modern. Thus, internet technology change any countries' travel agents or tour operators' air ticket sale method. So, it brings electronic ticket sale method is more popular to compare to traditional travel paper air ticket sale method.

However, online electronic ticket sale method has its disadvantages such as online transaction is unsafe, if the consumer 's name and address and visa card number is stolen to let any internet users to know to be used to buy any products from internet channel easily. Otherwise, traditional walk in offline travel paper ticket sale method is more sfe, because the travel consumers can pay cash to the travel agents directly.

However, offline travel agent disadvantages include that the research identified that information communication and technology has very crucial role for tourism industry. Tourist can access any kind of information about tourism destination and tourism products from any part of the world. Tourism comprehends with social media. For example, it was found that (ICT) is bosting up tourism industry. (ICT) helps in searching the location, search for information on tourism products, and e-booking of airline tickets and hotel reservation.

The online travel sale service attraction is that the recent development in the field of informatin communication and technology and its practical application in tourism and hospitality industry. Generally , online travel sale service must have consumer side and the supplier side.

The decision making prcess of consumer was analyzed and it was found that travel information search and traveller individual electronic ticker pre paid to prebook any plane seat, hotel rooms and restaurants prices comparison to prebook service of traveler individual purchase behavior are corresponding with the usae of (ICT).

1.4 What is the online travel sale service strategy?

The two most important things for travel operators (agents) are online travel marketing and strategic management. Former can enhance business operations. Use of (ICT) develops financial capabilities , however, it depends on management choice, financial condition and position. Some researchers recommended that the usage of IT should not be restricted at operatinal level, however it should be extended up to senior level and should be used for decision making. Social media is regarded as a platform where the tourists and travel operators/agents (suppliers) of tourism industry cross each other. Thus, the role of social media has been directed for future research in tourism industry. Hence, it seems online travel sale service has these features to attract travel consumers to choose to use this online mode to buy electronic air ticket. Such as, airline electronic air ticket price comparison, pre-booking plan seats to avoid full seats flights to delay consumer individual trip plan, pre-booking hotel rooms and prices comparison as well as prebooking restaurant seats and food price and taste comparison, travel destination easy search. Otherwise, these features to attract travel consumers to choose to walk in to travel agents to buy paper air ticket directly. They include: safe cash or visa card payment to avoid personal information is stolen by website payment channel, e.g. via card number, address, name , birth date personal information. Also the travel consumer can enquire any questions from the travel agent and gets individual feedback from the travel agent by oral before who ensure to choose to buy which kind of travel package for whose travel destination. In special, when the travel consumer has much time to spend to enquire any travel trip question, walk in travel agent is the best enquire methos to let the travel consumer to know the trip information clearly.

Online/offline travel operators
(agents) maketing strategies

2.1 Offline walk in travel unique segment service strategy

Nowadays, online and offlce travel operators competitions are serious. In fact, tourism marketing , there will be more need for online travel operators in the future, due to online travel sale service is popular to be accepted by online travel consumers. Thus, I recommend walk in offline travel agents need to concentrate on focusing some unique travel service to attract new or old travel consumers if who hope to survive.

I recommend that they can focus on specific specialized services, such as travel consultation (specialization) hypothesizing that systematic differences exist between the usage of travel agents for different travel contexts and travel agents can survive if they focus on specific segments of the market, such as older travelers (segmentation; hypothesizing that systematic differences exist between the usage of travel agents depending on the personal characteristics of travellers). The unique travel needs include: specific services related to package holidays, transport services, beach on city holidays, as well as destinations travellers are not familiar with.

I shall give my opinions to provide insight into alternative strategies for travel agencies in a matured travel market with a high internet penetration as below:

The internet online travel sale service is a reality of popular to let travel consumers to feel convenient to pre-book air seat, hotel rooms , air electronic ticket prices comparison. In order to make final purchase decision very easily in the shortest time. Consequently , it has penetrated the decision making process of travel to attract them to choose to buy electronic air ticket, prebooking hotel rooms or restaurant seats from online travel agent channel more than walk in offline travel agent channel. This is especially true in the tousism business where consumption to consume (booking) and the purchase-related information search (Bieger & Lasesser 2004; Crotts 1998).

In fact , apply website to provide travel sale method has these good consequence. From travel operator (agent) supplier's perspective, the success potential derived from operating a website consist of lower distribution costs, higher revenues and a larger potential market share (due to the ubiquitous access). From traverler's perspective, the internet allows direct communication with tourism suppliers facilitatinf requests for information and allowing services and travel related products, e.g. prebooking hotel rooms, restaurant seats , electronic or paper air tickets, travel trip arrangement package products to be purchased at any time and any place from online travel agents /operators conveniently.

Offline / online travel agency (operator) business depends on earn commissions on behalf of airlines. Thus, offline walk in travel agency (operator) business model that would extend existence as a booking agency (thus focusing on consultation and interpersonal contact) strategy.

As a matter of fact, commission -cutting , which began in the US well ahed of Europe, has had a profound effect specially on business travel agents . Consequently , many of them have re-invented themselves as " travel managers", instead of selling tickets and making arrangements, they charge consultancy fees for reducing the amounts client companies spend on travel (Daneshku, 1999).

2.2 Systematic differences strategy applies to offline walk in travel agent

Thus, I recommend systematic differences strategy can be applied offline walk in travel agent (operator). It means that walk in travel agents could reorient their offline walk in travel agent business to focus on contexts that are less substitutable by other channels and media . Factors hypothetically attributing to the delineation of travel contexts include: helping travellers to choose best travel destinations, helping travellers to attempt to find the number of previous trips (indicating the familiarity with a destination) for their travel reference, helping them to find the cheapest, the most convenient and the most close transportation to ctch during their trips, helping them to find the different types of accommodation and rooms price comparison , nature/type of the trip comparison , arrangement of time of booking (as indicator of spontneous / planned travel) nd helping them to budget overall travel expenditure .

Systematic differences in travel agent use exist in dependence of personal (characteristics with with tourists. Walk in offline travel agents could benefit from a travelling client segmentation strategy and customize and target their services to those travellers that are most likely to be and remain their customers.

Factors hypotheticlly attributing to the traveller segment include: travel expenditure per day, useful travel information as indicator for perceived risk and socio-demographic (age, gender, highest completed and education, professional positions) . Generally, the role of walk in offline travel agent with regard to the travel infrormation search and booking behavior have take an incoming perspective. Such as looking at visitors from different travel markets at a similar destinations. The comparison of central importance in determining whether specialization of travel contexts or market segments is the more promising strategy for walk in offline travel agents.

However, travel package tours strategy must b offline walk in travel attraction . Due to some walk in travellers target segmentation market has still needs. Generally, this travel package tours of travel segmentation consumer who like to enquire the travel agents to concern what the hotel rooms price are the cheapest to provide to them to live, what transportation tools the travel agent can arrange to them to catch anywhere the country destination, the travel agent can provide them to visit during their tour journey. Thus, the travel trip package service is still popular need to offline walk in travel agent (operator). This market is only belonged to offline walk in travel agents (operators) nowadays.

2.3 Service fees and commission cuts strategy

The reduction or removal of airline commission continues to challenge travel agencies' profitabilityl It is crucial to understand what trends travel agencies need to be aware of to ensure how to optmse profitability and increase travel agencies‘ revenues with service-fee models.

Service fees are not only a way to compensate for the loss of airline commission but also a way to generate new revenue sources for travel agencies that guarantee their long term profitability. Many travel agencies are expanding their service fee models, both in terms of the mounts changed and the number of service to airline.

However, if travel agent charge too much service fee to exceed the general airline travel market service fee reasonable or standard level. It will influence many airlines do not choose to find the travel agent to help them to sell air tickets. Travel agents apply fees most often for airline related services. They charge differentiated fees depending on the destination, type of reservation (e.g. frequent flyer), number of tickets sold or type of airline (e.g. full service versus).

However, service fee increaes can raise customer loyalty and satisfaction. It won't reduce client numbers or result ina losee in clients.. The reason is that service fees can be tailored to suit individual customer. Ths helps travel agencies target their clients, with tailored services based on their past purchasing patterns and identity services for which clients' willingness to pay is greater , such as trip planning identity service for which pay , such as hotel only or special promotion.

To revenue mix for travel agencies is increasingly shifting to service fes as airlines have lowered or cut commissions. Successful travel agencies in many European countries are fast adopting, and constantly upgrading , their service fee schemes. Thus, it seems reasonable service fee level is one important factor to influence travel agents and airlines good relationship. In fact, even travel agents raise service fee, it won't influence travel consumer number to be reduced , even they raise air ticket price. It they can provide the informations concerning the reasonable hotel roomes prices and food quality comparison to satisfy travel consumers' living arrangement or helping them to find the reasonable restaurants' food prices and where are their location arrangement or providing the reasonable airlines' electronic air tickets or paper air tickets sale service, even arrangement any high entertainment quality of travel destination trips to let travel consumers to feel satisfactory.

However, I believe the raise air ticket price factor won't inflluence the travel consumer number to be decreased. Any offline or online travel agents will encounter this crisis. By cutting travel agents' commission. Airlines decreased their dependence on travel agencies as a distribution channel. In fact, three key variable factors will influence travel agents' commission income to be decreased. They include below:

- The unsustainable or no change financial losses by airlines , due to the growth of low cost carriers, leading to an increase in the number of bankruptcies.
- No negative consequences from previous commission cuts: airline had progressively lowed the commission payments.
- No effective resource for travel agencies to satisfy airlines needs.
- The appearance of now airlines and air routes to provide to travel agencies to fall down air ticket price to attract consumers' choices,

due to who don't feel to spend much money to go to this new air routes or catch new airline plans , whether these new air routes are excite to entertainment or whether they are safe planes to catch.

- An increase in the number of bankruptcies to cause travel comsumption desire to be reduced.
- New competition forced down air fares.
- The necessity to cut production costs, especially with low cost meaning low production costs and low fares, even if the two are closely linked.

2.4 Internet negative influences to travel agents

Although, on the one hand, internet creates offline travel agents to use websites to help them to sell electronic air ticket or travel related products, such as prebooking hotel rooms , restaurants, transportation tools etc. travel service. However, on the other hand, internet also brings travel agencies competitive disadvantage with regad to suppliers' direct websites , when airlines are able to control seat availability and prices. Indeed internet cause the decision is made by the airlines to reduce and/or eliminate travel agency commission has led them to use technology that many of their distrust or are not inclined to use, and to compare prices and travel schedules constantly.

As a result of this travel sale service environment, traditional offline travel agencies are at a competitive disadvantage with regard to online travel agencie and to airline carriers, which have developed their own direct websites where they are able to control seat availability and prices.

Nevertheless, travel agents' pay programmes remain. From some airlines, travel agents receive negotiated incentive commission closely linked to their performance as incentive . However, airlines still need travel agents' assistance to help them to promote air tickets to sell, due to travel agents can provide trip packages, transportation tools, prebooking hotel rooms, restaurants and air tickets arrangement and they can give any enquiries to every individual travel consumer. It is free charge travel professional

enquiry service for travel agency's competitive features.
Consequently, how agencies can reduce their reliance on airline commission payments. I recommend these following strategic options to them to apply as below:

- Streamlining operations, controlling staff costs, when ensuring the client feels as little impact as possible.
- Expanding or moving into the leisure business, where commissions on ono-air products remain high (cruise, hotel, railway travel)
- Specializing in geographic areas or becoming niche players for specific leisure products, e.g. destination weddings, student travel group cultural travel, cruises only, cruise and railway travel etc.
- (d) establishing a service fee driven business model.

2.5 Concentrating on business travel marketing strategy

The certain characteristics to the business travel market allowed this sector to adapt more easily to the disappearance of commission. Business travel systems have always had different relationship with different customers. They usually have long term buyer relationships, set up long before the commission cap. Some of them quickly renegotiated their contracts to include a transaction or management fee, knowing that the majority of these fee arrangements are specific the need of the client.

The reasons why airlines reduce commission to paid to travel agents. They include petrol costs increasing, e.g. indirect and by pass the established distribution chain by developing airlines' their own websites; reducing or removing commission paid to travel agencies. Consequently, the decision to cut travel agencies' commission clearly shows that airlines wanted to decrease their reliance and dependence on travel agencies as a distribution channel. Thus, the internet appears to be an efficient and cost-effective distribution channel. Also, by creating airlines' own websites and setting directly to their clients, airlines are also to control seat availability to their clients and prices to their websites.

2.5.1 What an e-commerce strategy is used by internet travel websites?

Nowadays, the commercial use of electronic travel ticket travel is common, the most purchased online products include, for example, the name brands in online travel Epedia.travel .com and cheap tickets have been or are being integrated in large online travel firms.

Generally, online travel websites apply these strategies to attract travel consumers as below:

Firstly, shopping mall strategy, means to conduct a comprehensive factors for e-commerce. The online service provider needs to organize catalogs of services, take orders through their websites, accept payments securely, send service or related document, such as airline tickets to consumers and manage client data , such as client profiles.

Secondly, portal strategy, portal websites , such as yahoo give visitors the chance to find almost everything , they are working for in one place. Websites , such as Altavista.com and yahoo.com provide users with a shopping page that links them to many sites carrying a variety of products. Once a client is familiar with a website, who will be more likely to use the online service.

Thirdly, pricing strategy, low price is as a major competitive weapon. It includes a comparison pricing on discount price or price negotiation to let online travel consumers to get the best electronic travel ticket price choice to buy any airline tickets.

2.5.2 Travel agents vs online booking: Tackling the shortcomings and strengths

Consequently, however, one travel consumer who chooses either online booking sale service or traditional walk in offline travel agent to enquire travel service. These both of travel sale methods have shortcomings also. Such as it is possible that online electronic travel ticket purchase has personal data ,e.g. visa card, name, birth data, address, which will be stolen by online crime internet users more easily, who can not enquire any travel questions to get clear travel information concern whose travel destination package service choice or hotel room choice or transportation tool or restaurant choice and airline choice by travel agent. Also, it is possible that walk in travel agent paper travel ticket purchase shortcomings include that the travel consumer can not check any airlines' seat and pre book hotel room or transport tool or restaurant in the shorten time if who needs to fly immediately. Thus, it seems that online travel agent's client group is business travel intention, who does not need to enquire travel agent and has desire to per book airline seat in the short time. Otherwise, the offline walk in agent's client group is entertainment intention , who need to walk in to travel agent to enquire whose travel package and has no desire to pre book airline seat in the short time. Thus, online travel agent ought concentrate on design good travel package for the business travel consumers. Otherwise, offline travel agent ought concentrate on design good travel package for the entertainment travel consumers. Thus, they can have themselves unique travel target package to adopt to their different travel need. Such as business travel consumers need to live cheap and comfortable hotels, catching cheap and fast transportation tools in their business trips, eating in cheap and good taste food in restaurant and spending the less time to catch the airline plan to arrive the destination and cheap and comfortable business class plan seat. Such as entertainment travel consumers need the travel agent can help them to design cheap and enjoyable travel package, includes living comfortable hotel room, exciting and enjoyable trip, good taste food and

railway, travel bus, cruise and plane provision in trip.

In conclusion, In fact, tourism is a quite unique area of business in a sense that is a travel sale service product and it can't be observed or manipulated through direct experience prior to purchase . Instead clients have to purely rely on indirect or virtual experience. Thus, every online or offline travel agent ought attempt to design different travel package to attract every business traveler or entertainment traveller trip need because every traveler will have personal unique trip need in this competitive travel sale service market in the future.

Reference

Bieger. Th., and Ch. Laesser (2004). " Information sources for travel decisions: Toward a source process model," Journal of travel reserch, 42(4): 357-371.

Daneshku, S. (1999). " Unwived travel agents unworried bi internet, " Financial Times , London. June 16, 1999:10.

Foucault, B. Lery, N. Rifkin, A. & Silfies , 2000. “ Comparision of textbook prices by retailer and by college” working paper. Cornell University, Ithaca, Ney.

Chapter 3

III

Robots how influence human job behavioral and organizational mind change

(AI) -driven automation
industry development

1.1 (AI) - driven automation industry development how to influence work nature change to bring negative impaction

(AI) -driven automation industry will create wealth and expand economy growth to any countries, but it will be accompanied by changed in the skills that workers need to learn. One of main ways that technology increases productivity is by decreasing the number of labor hours needed to create a unit of output. It implies (AI) technology will influence low educated and low skillful labor number to be decreased (reduction employment number).

Although, AI automation can raise efficiency and productivity, but due to manufactuers will concentrate on bring (AI) robotic automation skills to manufacture any products, it will lead future many human jobs disappeared, finally, it will bring negative impact to bring many low skillful labours will lose their jobs , due to robots can replace them to do different kinds of jobs. (AI) -driven automation industry will create wealth and expand economy growth to any countries, but it will be accompanied by changed in the skills that workers need to learn. One of main ways that technology increases productivity is by decreasing the number of labor hours needed to create a unit of output. It implies (AI) technology will influence low educated and low skillful labor number to be decreased (reduction employment number).

In contrast, technological change tended to work in a different direction throughout the nowadays. The advance of computer and the internet raised the relative productivity of higher skilled workers. So, routine-intensive occupations that focused on predictable tasks disappearance, such as switch board, operators, filming checkers, travel agents and assembling line workers etc. were particularly replaced by new technologies. All these human jobs will be replaced from AI automation technology after ten years in possible.

In positively view point, (AI) driven-automation will make many workers more productive and increase demand for certain skills. Consequently, new jobs are likely to be directly create in areas , such as the development and supervision of (AI) as well as indirectly created in a range of areas throughout the economy as higher incomes lead to expanded demand. So, it is only the high skilled labors who can learn how to supervise or manage or control (AI) robots how to work in order to raise productive efficiency , whose skills will increase need. Due to their skills shortage, so they will be difficult to be replaced from (AI). It means that this high skillful (AI) supervision labours won't lose jobs in the future. Otherwise, they will raise wage, due to skill shortage in the (AI) industry job market future.

However, today, it may be challenging to predict exactly which jobs will be most immediately affected by (AI) driven-automation. The reason is because (AI) is not a single technology, but rather a collection of technologies that are felt unevenly through the economy to influence job changing both negatively and positively. In positively view point, (AI) driven-automation will make many workers more productive and increase demand for certain skills. Consequently, new jobs are likely to be directly create in areas , such as the development and supervision of (AI) as well as indirectly created in a range of areas throughout the economy as higher incomes lead to expanded demand. Otherwise, in negatively view point, many traditional human needed (demand) skillful jobs will be threatened by automation are highly concentrated among lower-paid, lower-skilled and less -educated workers. It means automation will cause pressure on demand for this group, pressure and employment, if (AI) can replace the low skilled and less educated workers' jobs. Thus, (AI) will have negative influence to impact on the labor market.

(AI) capabilities will enable automation of some tasks that have long required human labor. Why can (AI) replace some simple human jobs? For example, advances in robotics are expanding machines' abilities to interact with and sharp the physical world. Combined , (AI) and robotics will give rise to smarter machines that can perform more sophisticated functions than ever before and brings more advantages that humans have exercised. This will permit automation of many tasks now performed by human workers and could change the shape of the labor market and human activity.

1.2 How (AI) influences labor market

Today, it may be challenging to predict exactly which jobs will be most immediately affected by (AI)-driven automation. Because (AI) is not a single technology, but rather a collection of technologies that are applied to specific tasks.

Some specific predictions are possible based on the current (AI) technology. For example, driving jobs and house cleaning jobs, bank counter service jobs, telephone enquiry service operators.

Restaurant cooking jobs, simple accounting record service jobs etc. that require relatively less education to perform. Advancements in computer vision and related technologies have made the feasibility of fully appear more likely, potentially displacing some workers in driving-dominant professions. Seemingly similar robot, for which the operational tasks is less specific of navigating to a specific destination when following a set of given rules and preserving safety.

In the future, the effects of (AI) on the labor market in the decade ahead will continue the trend toward skill-biased change that computerization and communication innovations have driven in recent decades. Thus, some human driving occupation will be disappeared or replaced by (AI) automation driven. For example, bus drivers, light truck or delivery services drivers, heavy and tractor-trailer truck drivers, school drivers, tax drivers, travel bus drivers.

However, (AI) technology could enable some workers to focus time on other job responsibilities, boosting their productivity, and actually raised wage growth among those still holding the reshaped jobs. For example, salespeople, who currently spend a considerable amount of time driving could find themselves able to do other work when a car drives them from place to place, or inspectors and appraisers could fill out paperwork, when their car drives itself. This (AI) -driven technology should make these workers more productive, with (AI) -driven technology serving as a complement, not a substitute. New jobs will also likely be created, both in existing occupations cheaper transportation costs with lower prices and increase demand for products and all the related occupations, such as service and fulfillment, and in new occupations not currently foreseeable.

What kind of jobs will be created by (AI) technology? Predicting future job growth is extremely difficult, due to it depends on technologies or substitute for existing today as well as they may complement or substitute for existing human skills and jobs. However, (AI) will also lead to substantial indirect job creation to the degree it raises productivity and wages, it may also lead to higher consumption that would support additional jobs from high-end draft production to restaurant and retail. The future(AI) " augmented intelligence", the technology's role is as assisting and expanding the productivity of individuals rather than replacing human work. Thus, based on the biased-technical change framework, demand for labor will likely increase the most in the areas where humans complement (AI) automation technologies. For example, (AI) technology , such as IBM's Watson may improve early detection of some cancers or other illnesses, but a human healthcare professional is needed to work with patients to understand and translate patients' symptoms, inform patients of treatment options, and guide patients through treatment plans. Shipping companies may also partner workers who pick up and deliver products over the last feet with (AI) enabled autonomous vehicles that move workers efficiently from site to site. In such cases, (AI) augments what a human is able to do and allows individuals to either be move effective in their specially task or to operate on a larger scale. Thus, it seems (AI) technology will also create new jobs, raise productivities and workers' efficiencies.

1.3 Redefining management in
the workforce of artificial intelligence

- Change management

In the future, due to artificial intelligence influences to some kind of human jobs nature. So, the kind of human jobs of management methods will also need to change to adapt the artificial intelligence technology input to their organizations. It will cause challenges for every executive and manager if who won't have effort to manage their teams how to apply artificial intelligence technology to work efficiently and easily. For example, division of labor will change among humans and machines will increase. Thus, companies will have to adapt their training performance and talent strategies how to emphasize on work that how to make human judgment and skills and experimentation. Thus, (IA)'s greatest impact will be on administrative coordination and control tasks, such as scheduling , resource allocation.

In fact, mangers will encounter this challenges: How to apply human experience and expertise to judge critical business decisions and practices when the information available is insufficient to suggest a successful course of action? Due to this kind of work will require new skills and mindsets. I shall indicate these change management methods to adapt (AI) technology. Such as: administration and routine tasks, scheduling , allocation of resources and reporting will fall within the intelligence machines, responsibilities that have long been reserved for humans. For example, a typical store manager or a lead nurse at a nursing home most constantly arrange shift schedules, accounting for staff members' absences owing to illness, vacation time or sudden departures.

Thus, the managers need to learn how to arrange new division of labor within the organizations after (AI) technology had been implemented to the organization. Artificial intelligence is currently influencing into once considered exclusive to humans: assessing and acting on human emotions and personality traits. The influences to managers need to change their strategies to adapt (AI) technology implements include such as below:

Firstly, managers need to spend the bulk of their time on coordination and control tasks from intelligent system implements. Their time spending on these major three aspects from impact of intelligent system: coordinate and control, solve problems and collaborate and people and community , strategy and innovation three aspects. Thus (AI) will influence managers need to change their judgment method to teach whose teams how to adapt the (AI) system operations in any organizations.

Secondly, (AI) will influence top, middle and low level management needs to change to adapt the (AI) technology operations to any owned (AI) technology organizations in the future. Intelligent machines must be trained in context. Just like humans , on-the-job training is a requirement for such machines because they typically arrive with only very general capabilities. To get the most from (AI), managers at all levels must participate in the instructional experience and in the learning process and provides managers' familiarity with such systems on these aspects, e.g. How the system works and generate advice, how the system has a proven track record , how the system provides convincing explanations , how the system can make simple rule- based decisions.

Thirdly, managers need to learn how to make judgment more accurate (AI) systems assistance. Although (AI) will invariably take on more routine work and even augment human decision-making, it won't judgment work, the application of human experience and expertise to critical business decisions when the information available is insufficient to suggest a successful course of action or reliable enough to suggest an obvious course of action. For a sense of the nature of judgment work, consider big data marketing and sales analytics. Such analytics often provide insights that can inform promotional campaigns, including predicting which promotions will generate desired sales brand further into the future, marketing executives need use judgment, combining analytics with their own and others' insight and experience.

The application of experience and expertise to critical business decisions and practice represents the real value of human judgment. But, when artificial intelligent machines are implemented to any organizations to assist the low, middle and top level management to make any business judgment. These forms of judgment work that managers can gather data interpretation, idea development more absolute from (AI) machine assistance. Thus, why these level management executives need to learn how to apply (AI) machines to help them to make any business judgment more accurate.

- How (AI) influences organizational change

Consequently creative and social intelligence will be in even greater demand as (AI) makes in management and the workforce. This development will represent a long term trend in labor markets , one characterized by intensifying demand and reward for social skills with a growing desire for creative capabilities, managers will seek to fashion of ideas and hypotheses from inside and outside of the enterprise to shape solutions to their most pressing business problems. Thus, (AI) will influence overall organizational team members who have chance to participate any decision to make more accurate business judgment.

Many managers mistakenly view judgment work as only an individual discipline, failing to appreciate that it can also involve decide interpersonal and organizational practices. In more complex settings, judgment is typically a collective outcome of individuals' and teams' diverse perspectives, insights and experiences. And often , the resulting choices are better informed than decisions that an individual would have arrived at on his or her own.

Thus, when any organizations apply (AI) technology to assist managers to gather data and ideas to make any judgment. In these cases, organizations can create the conditions for effective collective judgment by establishing structures , such as " shadow advisory boards" that prompt managers and employees to source and synthesize multiple perspectives. Thus, a traditional organization (firm) might freshen its thinking is t put together a shadow advisory board, comprised of young, digital people who can apply (AI) machine assistance to make judgment work more accurate whether related to people development, problem-solving or strategizing and innovating for considerable degrees of creative and social intelligence.

Thus, on the one hand, (AI) technology machine augmentation and automation can give these advantages to human (organization managers) , e.g. developing people and community, solving problems and collaborating, coordinating and controlling work, shaping strategy and leading innovation. Besides, on the other hand, the next generation managers need have these individual attitude to treat intelligent machines to be as colleagues.

When, judgment is a human skill, intelligent machines can accelerate human learning that supports it, assisting in data -driven simulations, scenarios and search and discovery activities. Focuses on judgment work, some decisions require insight beyond what data can tell them. This is the sweet sport for human judgment, the application of experience and expertise to critical business decisions and practices. Thus, managers will also need to find ways to learn how to use digital (AI) technologies to tap into the knowledge and judgment of partners, customer external stakeholders and role models in other industries after the (AI) machine had been implemented to the organization.

Future works change:Automation, employment and productivity

2.1 How (AI) influences employment

Human future " micro to macro" industry trends will be affected business strategy and public policy by (AI) technology. In the future (AI) technology will influence those six themes: productivity and growth, natural resources, labor markets, the evolution of global financial markets, the economic impact of technology and innovation and urbanization. However, (AI) technology will bring economic benefits of tackling gender inequality, a new global competition, Chinese innovation and digital globalization.

Nowadays, advances in robotics artificial intelligence, and machine learning are in a new age of automation, as machines match or outperform human performance in a development to any countries. For example, automation of activities can enable businesses to improve performance by reducing errors and improving quality and speed, and in some cases achieving outcomes that go beyond human capabilities. For example, some research indicated automation could raise productivity growth globally by 0.8 to 1.4 % annually; more than 2,000 work activities across 800 occupations. When less than 5% of all occupations can be automated using demonstrated technologies about 60% of all occupations have at least 30% of constituent activities that could be automated. Many occupations will change that will be automated away: Activities most susceptible to automation involve physical activities, in highly structured and predictable environments, as well as the collection and processing of data. They are most prevalent in manufacturing , accommodation and food service and retail trade and include some middle-skill jobs. For example, such as natural language processing is a key factor. Beyond technical feasibility, the cost of technology competition with labor including skills and supply and demand dynamics, performance benefits including and beyond labor cost savings, and social and regulatory acceptance will be affected by (AI) automation technology. Thus, (AI) automation will impact to influence global employment in those aspects as below:

Firstly, assuming that people are displaced by automation will find other employment. The anticipated shift in the activities in the labor force is of a similar order as the long-term shift away from agriculture and decreases in manufacturing share of employment. Both of manufacturing and agriculture industries which would be accompanied by the creation of new types of work not foreseen at the time.

Secondly, for business, the performance benefits of automation are relatively clear. Thus, the businessmen have opportunities for their micro economies to benefits from the productivity growth potential and macro economies to benefit to encourage continued progress and innovation , investment and market incentives. At the same time, employers must innovate policies to help workers and institutions adapt to the impact on employment.

This will likely include rethinking education and training, income support and safety nets , as well as support for those dislocated, when employees need to leave themselves homes to move to other cities to learn new (AI) automation works. Thus, individuals in the workplace will need to engage move comprehensively with machines as part of their everyday activities, and acquire new skills that will be in demand in the new automation age. Consequently , the scale of shifts in the labor force over many decades that automation technologies can be a similar order to the long -term technology -enables shifts in the developed countries' workforces away from agriculture in the 21 th century. Those shifts did not result in long-term mass unemployment because they were accompanied by the creation of new types of work not foreseen at the time. However, human will still be needed in the workforce when the total productivity gains are caused by (AI) technology.

2.2 What occupations will be influenced by (AI) technology.

In the future, scientists predict that these occupations will be influenced by (AI) technology mostly. They include : retail salespeople, food and beverage service workers, language or translation teachers, health practitioners. Since these work activities have a more relevant occupations are made up of a range of activities with different potential for (AI) automation . For example, a retail salesperson will spend more time interacting with customers, stocking shelves , or ringing up sales. Each of these activities is distinct and requires different capabilities to perform successfully.

Thus, these job activities have similar simple control characteristics. Simple activities include greet customers, answer questions about products and services, clean and maintain work areas, demonstrate product feature process sales and transactions. All these activities can have similar simple activities in order to (AI) machines can be learn how to do these activities from (AI) technology . For example, the capability perception includes sensory perception, cognitive capabilities, such as retrieving automation, recognizing known patterns(supervised learning), logical reasoning problem solving.

Thus, (AI) machine is such human, which has feeling and emotion, such as social and emotional sensing, judgement reasoning methods, natural language understanding and physical capabilities, such as mobility , navigation, gross motor skill, fine motor skills. It seems that the future, (AI) human invents machines which will have these human characteristics to do human similar behavioral job duties more easily and efficiently. It implies these above human occupations will be replaced by (AI) human invention machines in the future. Due to (AI) creation, it is possible to cause unemployment number of these above workers will increase because (AI) machines can do their similar job behavioral activities. Consequently, employers won't need to employ many of these skillful labor. Otherwise, they can buy less number (AI) machines to attempt to do whose job activities more easily and efficiently. So, it seems (AI) machines will have more high work performance to replace these occupation workers' work performance. Finally, these occupation worker unemployment number will only increase when the (AI) machines had been invented to achieve to do their work behavioral activities absolutely success in the future.

2.3 Whether (A) technology machine labor will replace human worker more or assist human worker more

There is no single agreed definition of a robot how outcome of a task that is completed without human intervention. When some definitions require the task to be completed by a physical machine moves and respond to its environment, other definitions use the term robot in connection with tasks completed by software , without physical embodiment.

However, to answer the question : Whether (AI) technology machine labor will replace human worker more or assist human worker more. I shall indicate some examples to let readers to judge whether (AI) technology can create new jobs or reduce old jobs.

Firstly, I shall explain what (AI) function is. (AI) is a service robot that performs useful tasks for humans or equipment excluding industrial automation application . Thus, the classification of a robot into industrial robot or service robot is done according to its intended application. It is also a personal service robot or a service robot for personal used for a non commercial task, usually by lay persons . Examples are domestic servant robot, and pet exercising robot. It is also a professional service robot or a service robot for professional used for a commercial task, usually operated by a properly trained operator. Examples, are cleaning robot for public places, delivery robot in offices or hospitals, fire-fighting robot, rehabilitation robot and surgery robot in hospitals. Thus, these functions will be future (AI) application to our daily life necessaries

or business necessaries.
However, some authors agree (AI) will bring negative outcomes of automation, due to raise competiveness, reduce human job nature. Otherwise, other authors argue (AI) will bring positive outcomes of automation, due to raise productivities, job creation, assist humans work.
On the positive outcome hand, robots can increase productivity . This is particularly important for small-to medium sized businesses both are in developed and developing countries economies. It also enables large companies to increase their competitiveness through faster product development and delivery. Increased use of robot is also enabling companies in high cost countries to re shore, or bring back to their domestic base parts of the supply chain that will have previously outsourced to sources of cheaper labor. Currently , the greater threat to employment is not a automation, but an inability to remain competitive. Automation has led overall to an increase in labor demand and positive impact on wages. The reason is that the middle-income/middle-skilled jobs have reduced as a proportion of overall contribution to employment and earnings leading to fears of increasing income inequality, the skills range within the middle income bracket is large. Thus, robots are driving an increase in demand for workers at the higher -skilled and with a positive impact on wages. This issue is how to enable middle-income earners in the lower-income range to unskilled or retain. Finally, the (AI) positive impact supporter who argue the future will be robots and humans can work together.
However, on the negative outcome hand, robots can substitute labor activities, but don't replace jobs. They believe that less than 10% of jobs are fully automatable. Increasingly , robots are used to complement and augment labor activities, the net impact on jobs and the quality of work is positive. Automation can provide the opportunity for humans to focus on higher-skilled, higher-quality and higher-paid tasks. Robots can improve productivity when they are applied to tasks that which perform more efficiently and to a higher and more consistent level of quality than humans. For

example, increased productivity is enabling some firms, such as Whirlpool, Caterpillar and Ford Motors company in the US restructure their supply chains, bringing back parts of the manufacturing process to the country of origin. Thus, productivity gains due to robotics and automation are important not just at the company level, but also for build industry and nation competitiveness.

I suppose that productivity can be raised. What are the impacts of robots on employment? Firstly, the main focus of development has been on personal entertainment, which does not drive worker productivity (manufacturing production). When the internet (information and communication technology (ICT)) innovation. This is borne and by findings that manufacturing productivity, which has been driven by innovations in automation rather than consumer technologies, has government strongly than productivity in the services sectors of the economy in most nature economies. It seems (AI) automation will create many jobs in internet communication entertainment game industry. For example, many young people like to use internet to play any electronic games from computer or mobile at home or outside home conveniently. Thus, (AI) automation will increase demand to be invented to any new entertainment game from internet channel. It will need to employ many (AI) entertainment game inventors to create many automation entertainment games. Thus, (AI) automation in internet entertainment game industry will need human (AI) entertainment game inventors to invent the knowledge-based capital of (AI) automation entertainment games. The (AI) entertainment game inventors will need own research and development skills, form specific skills, organizational know-how skills, databased knowledge, design and various forms of intellectual property to do these (AI) automation entertainment game invention occupations in the future.

International Federation Of Robotics(2016) indicated that China will be as a major robotics manufacturer and user of robots, benefiting from jobs created by robot manufacturing and productivity gains from robot use. Chins had sold of robots to any one single market every year since 2017 year. The Chinese government has included a focus on robotics in its 10 year strategy. In order to achieve its target of a robot density of 150 units per 10, 000 workers by 2020 year. Thus, Chinese companies will have to install around 650,000 new industrial robots between 2016 to 2020 year, 2.5 times more than installed globally in 2015 year.

Hence, China (AI) manufacturing industry will need to employ many workers . It implies (AI) manufacturing industry will create many new occupations in China. Also, ministry of economy, trade and industry (2015) also showed that Japan currently has the largest stock of industrial robots in operations, primarily in the automation industry. Driven by a rapidly aging population and low productivity rates, the Japanese government has sights on a 20-fold increase in the use of robots in the non-manufacturing sector and a three-fold growth rate of labor productivity in the service sector both by 2020 year. Thus, it also implies Japan will need many robots to be provide to service industry. Due to robots will provide to serve any businessmen's clients. Thus, it is possible that the service workers won't be dismissed as well as it is depended on the serving job nature to decide whether Japan's service workers can still serve to their employer when the service (AI) robots are applied to whose employers.

Consequently, it seems that (AI) can create employment, Ministry of economy, trade and industry (2015) showed that such as China will develop the major (AI) automation manufacturing industry. The (AI) employers will need to employ many workers to manufacture any these different kinds of (AI) robots to satisfy China or overseas individual or business buyers needs. But, (AI) can also cause unemployment to the low skillful service workers. Such as if Japan some service businesses choose to buy any (AI) service robots to replace their service staffs to serve their clients. It is possible that the service staffs will be dismissed, due to (AI) robots can do such as their same service job duties to achieve better service performance. Thus, today, it is increasingly common for people to use robots in various situations at home and in retail stores, hotels and hospitals these service industries. Robots are classified into server types based on their functionality (service and utility robots or those designed to communicate with humans) and appearance (humanoid robots or mechanical robots). The type of robot, to which each country allocated particular importance in the advance of robotics, reflects the sense of values and preferences of its population. Thus, if the country has high population needs to use robots, then they will influence either more new jobs creation or more old job loss in the country's (AI) manufacturing or (AI) service industries both. For example, Japan respondents often associate the term " robot " with humanoid robots that can communicate with human and they have a high level of familiarity with robot. The US has the highest level of robot utilization at home and in retail stores with its people being the most enthusiastic about the future use of robots. Germany shows a strong tendency to consider robots for industrial purposes and its people feel strong effort to the presence of robots in their households.

In conclusion, to judge whether how (AI) will influence the country's employment to be better or worse. It will depend on the country home buyers (users) or business buyers (users) how to use (AI) for their daily needs. If the country , such as US retail stores need to use (AI) , it will have possible to reduce some or many retail service workers. Even, if the country , such as Japan has many home users need to use (AI) , it will not influence the employment market. Otherwise, it will raise (AI) salespeople numbers. Even, if the country, such as Germany and China will have many (AI) manufacturers, then it will create many (AI) manufacturing occupations for these (AI) manufactory workers.

Consequently, (AI) robots manufacturing and service needs will have positive or negative impact to any country's employment. It will depend on the (AI) service provision and service workers' job nature as well as the manufacturing workers of (AI) knowledge level to decide their employment chance in their country's employment market.

AI brings what jobs change

What does artificial intelligence(AI) mean?

- What (AI) function is?

Some scientists explain that artificial intelligence means which is an expert system, computer software that embodies a portion of the specialized knowledge of a human portion in a specific, narrow domain, owns decision making ability of human expert. The (AI)technology is based on the premise that what makes a person an expert is years of experience that enables who recognizes certain patterns in a problem as being similar to pattern. For example, in the future artificial intelligence system can be applied to control air traffic, design to computer configuration, medical diagnosis, instruction/training, speech/interpretation, monitoring to (nuclear plant), planning to mission, factory scheduling, prediction weather, repairing telephone, automatic driving etc. different industries.

Artificial intelligence characteristics include: creative, adaptive , common sense, fact processing, quick replication, broad focus permanent and consistent skill. Otherwise, traditional computer expert system disadvantage includes perishable, unpredictable, slow reproduction, expensive, slow reproduction, slow processing lacks inspiration, needs instruction, narrow focus only machine knowledge. So, artificial intelligence is a branch of computer science devoted to creating computer to influence software and hardware to attempt to create human intelligence or human intelligent behavior. It is learning from experience, responds flexibility in situation that are, new or not anticipated.

Thus, (AI) can be learnt programmed knowledge to solve problems, using reasoning in solving problem, understanding and inferring facts and rules, recognizing the relative importance of different elements in a situation. In summary, artificial intelligence is concerned with two basic ideas mainly: The first idea, it involves studying the thought processes of humans to understand what intelligence is; the second idea, it deals with representing thought processes using companies to create artificially intelligent entities for testing the theories of intelligence.

- Can (AI) impact human job nature?

Human need concern this question: Will artificial intelligence (AI) reduce some human jobs in order to instead of replacing machines to do? Due to artificial intelligence is the ability of machines to do thing, that people would require intelligence. For example, artificial intelligence machine man driving(self-driver), it (AI) machine man driving research is an attempt to discover and describe aspects of human intelligence that can be simulated by driving machine functions. Alternatively, (AI) mathematical research may be another viewed as an attempt to develop a mathematical theory function to describe the abilities and actions of things (natural or man-made) exhibiting intelligent behavior and server as a design of intelligent calculation machine function.

Why do humans need artificial intelligence machines to instead of traditional human service job? For example, can artificial intelligence machine man (self-driving) driver drive to replace human driver? I shall compare the differences between humans and computers : The characteristics of humans are good at recognizing various things, either seen before or not, recognizing the relationship patterns between things. Human thinking is common sense reasoning, combining all types of sensory input, acting appropriately in novel situations, learning new things and changing behavior patterns, making decisions , even when given incomplete information, working with noisy, incomplete information gathering behaviors . However, characteristics of computers are good at: The tasks humans do naturally are extremely difficult for a computer program as intelligent, which must be able to do the same kind of tack as humans do naturally.

Hence, (AI) is an combination of many different success and technologies: Linguistics - computational and socio, philosophy-logic, philosophy of mind and of language, electronical engineering -image and speech processing, pattern recognition, robotics, machine learning, neural networks, optimization scheduling, management information system and decision making. So, it is possible that (AI) can impact human job nature to instead of human working behavior in the future.

How can (AI) influence labor market?

- How can human society job nature

to be changed to artificial intelligent society?

From the first intelligent perspective reason view point, artificial intelligence is making machines " intelligent" acting as humans expect people to act. Artificial intelligence has ability to distinguish computer responses from human responses, it owns knowledge to solve expert problem. From another research perspective reason view point, artificial intelligence is the study of how to make computers do things which, at the moment, people do better (Rich & Knight, 1991, p.3).

(AI) researchers are native in a variety of domains, e.g. formal tasks (mathematics, games), tasks (perception, robotics, natural language, common sense reasoning), expert tasks (financial analysis, medical diagnostics, engineering, scientific analysis and other areas).

From the second business perspective reason view point, (AI) is a set of many powerful tools, and methodologies for using those tools to solve business problems. From a programming perspective reason view point, (AI) includes the study of symbolic programming problem solving and search .

From the third human technological perspective reason view point, today's computer can do many well-defined tasks, for example, arithmetic operations, are much faster and more accurate than human beings. However, the computers‘ interaction with their environment is not very sophisticated yet. How can human test whether a computer has reached the general intelligence level of a human being? Can a computer convince a human interrogator that it is a human? But before thinking of such advanced kinds of machines, human will start developing our own extremely simple " intelligent" machines.

So, it is possible that human society job nature will to be changed to artificial intelligent society when (AI) technology is developed to the mature stage in the future.

- Why does human need artificial intelligence machines?

One of major division in (AI) is between humans who think (AI) is the only serious way of finding out how we (human) work and human who want companies to do very smart things, independently of how we (human) work. This is the important distinction between cognitive scientists vs engineers. One of another major division in (AI) is between symbolic (AI), which represents information through symbols and their relationships. Specific Algorithms are used to process these symbols to solve problems or deduce new knowledge and connectionist. So (AI) , which represents information in network. Biological processes underlying learning, task performance and problem solving are imitated from human mind behaviors. Thus, it is possible that artificial intelligence machines can do the better judgicious behavior to compare human.

- How does artificial intelligence influence future working changing in automation employment and productivity aspects?

In the automation changing influence aspect, as companies increasingly use robots on production lines or algorithms to optimize their logistics manage inventory, any carry out other core business functions. Technological advances are creating a new automation age in which ever-smarter and more flexible machines will be deployed on an ever larger scale in the marketplace. However, researching artificial intelligence with how influences human working nature. We need to answer these questions: How will automation transform the workplace? What will the implications for employment? And what is likely to be its impact both on productivity in the global economy and on employment?
Advances in robotics, artificial intelligence, and machine learning are growing in a new age of automation as machines match or outperform human performance in a range of work activities, including ones requiring cognitive capabilities. What factors are

determined the changing in workplace adoption by artificial intelligence innovation? What advantages are automation? Automation of activities can be enabled businesses to improve performance by reducing errors and improving quality and speed, and achieving outcomes that go beyond human capabilities.
Some scientists indicated based on their scenario modeling. They estimated automation could raise producing growth globally by 0.8 to 1.4 percent annually. Almost, the activities people are paid almost $16 trillion in wages to do in global economy have the potential to be automated by adopting currently demonstrated technology. According to their analysis of more than 2,000 work activities across 800 occupations. When less than 5% of all occupations have of least 30% of activities that could be automated. They also indicated that technical economic and social factors will determine automation. Continued technical progress, for example, in areas such as natural language processing is a key factor beyond technical feasibility , the cost of technology, competition with labor including skills, and supply and demand dynamics, performance benefits including and beyond labor cost savings and social and regulatory acceptance will affect (alter) the scope of automation.
Other some scientists also indicate U.S. country for example, the anticipate shift in the activities in labor force of a similar order of magnitude as the long term sight away from agriculture and decreases in manufacturing. Share of employment in the United States both which were achieved. So, those factors can influence why artificial intelligence technology needs. So, it is possible that future agriculture and manufacturing both industries will apply (AI) technology manufacturer-kind of job nature to raise productivity instead of farmers, fruit picking workers, farming transportation labours as well as factory manufacturing workers and supervisors etc. human-kind of job nature.

- Is artificial intelligence possible to replace labor ?

Not just intelligence, but also debating, if machines are capable of having a conscious minds. Artificial intelligence has those characteristics as below:

On functionalism aspect, artificial intelligence inputs mental states, sensory inputs, (beliefs, desires being in pain feeling) and behavioral outputs. Since mental states are identified by a functional role, which are thoughts to be manifested in various systems. Even, perhaps computers which are physical devices with electronic substrate that inform computations on inputs to give outputs similar to brains which are artificial intelligence composed of part any intrinsic relationship to each other. Thus, artificial intelligence activities is not the whole itself, but into parts or on external influence on the parts.

On dualism aspect, artificial intelligence is a set of views about the relationship between mind are matter. On materialism aspect, it builds the only thing that exists is matter, including consciousness. On biological naturalism aspect, it is similar a human brain than feels pains makes mental situation. So, artificial intelligence is similar biologist which might to be excited to human labor work. Hence, it seems artificial intelligence can change (alter) or replace human labor work of nature in possible in the future.

- Can (AI) technology replace human labour nature of work?

On technological innovation reason view point, the history development of artificial intelligence studying the intelligence is one of most ancient scientific discipline. The history development of artificial intelligence what aims to achieve human use to sense, learn remember and think, logic probability, decision making and calculation develop from mathematics, instead of replacement human labor functions.

Artificial intelligence history development aim is the scientific analysis of skills in connection and practice with the appearance of computers from 1950 year beginning. The artificial intelligence (AI) can deal with the ultimate challenges. How can (either biological or electronic) mind sense, understand and manipulate a world that is much simple and more complex than itself? And what if would human like to construct something with such capabilities?

The general-purpose software of the early period of (AI) were only able to solve simple tasks effectively and failed when which should be used in a wider range or an more difficult tasks. One of the sources of difficulty was that early software had very few or mix knowledge about the problems which handled, and activities successes by simply syntactic manipulation. Moreover, the other difficulty was that many problems that were tried to solve by the (AI) were untreatable.

The early (AI) software whether trying step sequences based on the basic facts about the problem that should be solved, experimented with different combinations till which found a solution. From the end the 1960 year, developing the so-called expert systems were emphasized. These systems had (sue-based) knowledge base about the field which handled. Till to the beginning of the 1970 year, (Prolog) the logical programming language was born, which was built in the computation realization of a version of the resolution calculus. (Prolog) is a remarkably prevalent tool in developing expert systems (on medical, judiciary and other scopes), but natural language parsers were implemented in this language. Then, in 1981 s, the Japanese announced the fifth generation computer system project, a 10 years plan to build an intelligent computer system that use the (Prolog) language as a machine code. Nowadays, (AI) can be applied any industries, such as car manufacturing industry can use (AI) technological machine-men manufacture car, instead of replacing human labors in factory. Even, in the future, using (AI) machine-men drivers can drive any private cars or public transportation tools, instead of replacing human drivers, e.g. bus, train, tram, ferry etc. Also in the future, machine-men can replace housewives to serve families to do housekeeping clean job , e.g. cleaning toilets, bathrooms, kitchens, even cooking functions at home. So (AI) machine-man can reduce housewives works at home. Moreover, (AI) machine man can take care old people , when who are living at homes or elder care centers.

So, it seems artificial intelligence (AI) will be possible developed to manufacture a new generation machine-man to assist (serve) families to do any simply cleaning or cooking jobs at homes. Moreover, the overall demand of (AI) general social needs will also rise, such as security, driving transportation tools, restaurant cleaning, elder centers care service etc. So, it seems that individual or families or social needs of (AI) will be increase in the future. Thus, it will influence macro economy growth (GDP) if there are large house family consumer group and hotel or bus or taxis or ferry etc. different business consumer group demand any artificial intelligence machine numbers increasing. Then, the artificial intelligence products and material manufacturers must need to buy many artificaial intelligence materials to produce any kinds of artificial intelligence machines to prepare to satisfy consumer individual needs. Consequently, macro economy will grow to the owned artificial intelligence development countries, e.g. US, China, UK.

● Why can artificial intelligence satisfy human needs?

First, On machine-man satisfactory demand aspect view point, it makes computers that think, it is the automation of activities. We associate with human thinking: like decision making, learning. It is the act of creating machine that perform function that require intelligence when performed by people. It is the study of mental faculties through the use of computational models. It is the study of computations that make it possible to perceive, reason and act. It is a branch of computer science that is concerned with the automation of intelligent behavior. It is anything in computing service that human don't yet know how to do property.

Second, on thought aspect artificial intelligence means systems thank think like humans, systems that think rationally.

Third, on behavioral aspect, artificial intelligence systems that act like human and that systems act rationally. However, the basic objective of (AI) is to represent human's thought processes in computation . These machines are supposed to exhibit behavior that. It is performed by a human being, would be considered intelligent. However, some authors feel (AI) has disadvantages, such as it is not creative, it is excited in the use of sensory devices, it can't make use of a very wide context of experiences and it does not use common sense.

For speech recognition and understanding function needs example, (AI) can be applied in speech recognition and understanding function, which (AI) speech or voice recognition is a data input method. For example, the computer recognizes and understands one (or a few) word commands. Speech understanding on the other hand is the computer's ability to understanding a spoken language. That is , the computer understands the meaning of sentences, an paragraphs through (AI).

So, (AI) can be attempted to learn human language how to speak. It is similar to translate human language skill, instead of actual human speaking skill. Also, (AI) can assist handicap learning or language student how to listen different languages by machine-man sounds from computers more accurately. So, it seems that it (AI) can replace human language teachers speaking function and can change teaching language nature of job in language speaking and listening education industry.

- Is artificial intelligence one good choice for human future technological benefit?

Nowadays, new technology development is popular. However, artificial intelligence is one kind of new technology choice among different technologies innovation. So it brings this question: Is artificial intelligence technology value to invest? To answer this question. I shall indicate some other new technology developments to compare (AI) technology development to judge which has urgent needs to achieve human expectation nowadays.

For example, why is green peace interested in new technologies? New technologies features prominently in our ongoing campaigns against genetic modified crops and number power. However, which are also an integral part of our solutions to environmental challenges, including renewable energy technologies, such as solar, wind and wave (water) power energy as well as waste treatment technologies, such as mechanical, biological treatment.

It seems humans need concern how to apply (AI) technology to solve environment pollution challenges in our future. So, environment protective, agriculture, natural energy technology will be popular demand to attempt to apply (AI) technology to solve their challenges or apply (AI) to assist to develop their industry.

What is influence to (AI) job nature change to change developed and countries countries economy

- How can artificial intelligence technology influence economy?

Advances in artificial intelligence (AI) technology and related fields have opened up new markets and new opportunities progress in critical areas, such as health, education, energy, economic development, social welfare and the environment pollution.

(AI) automation will continue to create wealth and expand the global economy development in the future. However, when many will benefits that growth won't be costless and will be accompanied by changes in the skills, that workers need to increase productivity in the economy and structural changes in the economy. So, in the skills that workers need to succeed in the economy and structural changes.

I shall indicate why aggressive policy action will be needed to help Americans who are disadvantaged by these changes , due to (AI) technology is caused. For automation industry change example, artificial intelligence (AI) capabilities will enable automation of some tasks that have long required human labor. These artificial intelligence technology introduction can increase new opportunities for individuals. The economy and society, but (AI) has also the potential to disrupt be current livelihoods of many Americans. However, (AI) leads to unemployment and increase in inequality over the long run depends not only on the (AI) technology itself, but also on the institutions and policies that are changed.

Thus, it is possible that (AI) technology will raise some countries unemployment number if the employer apply (AI) technology workers to work instead of human labor in their factories, but it can also raise productivities for these employers.

- Can (AI) influence global economy growth?

Technological progress is main driver of growth of GDP per capita, allowing output to increase faster than labor and capital . However, technology can increase productivity, but also decrease the number of labor hours needed to create a unit of output. So (AI) causes unequal to labor wage decreases, even reduces the number of labor to manufacture, e.g. artificial intelligence technology of automation car manufacturing industry; clothing manufacturing industry; plane manufacturing etc. high technology of artificial intelligence manufacturing method. But (AI) should be potential environment benefit, although it raises unemployment ratio. Moreover, it can rise production , due to many skilled craft were replaced by the combination of machines and lower-skilled labor. The result of (AI) technology introduction , it causes output per hour risen when inequality declined, driving up average living standards, but the labor of some high-skill workers was no longer as valuable in the market. Otherwise, if (AI) technology is continue developed to be success. Some routine intensive occupations will be loss, which focused on predictable, e.g. easily programmable tasks, such as switchboard operators, filing clerks, travel agents, and assembly line workers would be particularly replaced by new (AI) technology. However, at the same time, (AI) technology development will bring these benefits: improvement in education (training (AI) technology scientists) , due to (AI) manufacturing technology needs are raising to businesses and institutional changes, such as the reduction in unionization and raising in the minimum wage to the (AI) manufacturing technology skilled labor in factories.

Because (AI) technology is not a single technology, but rather a collection of technologies that are applied to specific tasks, the effects of (AI) will be felt unevenly though the economy. It will bring some tasks will be most easily automated than others , and some jobs will be affected more than others, both negatively and positively. Finally, new jobs are likely to be directly created in areas , such as the development and supervision of (AI) as well as indirectly created in a range areas though out the economy as higher incomes lead to expanded demand.

However, if (AI) technology could dominate global labor markets. If labor productivity increases, do not influence into wage increases, then the large economic gains brought about by (AI) technology could be increased wealth inequality, due to employers can reduce production cost, but workers (labors) wages will not be increased, even will be decreased. Hence, it seems the (AI) technology will bring disadvantages to labor market to cause unemployment or reduce wages in possible, although it can reduce employer individual salary (wage) expenditure and it can raise productivity.

- How can artificial intelligence impact global economy growth?

Artificial intelligence (AI) technology is a branch of computer science that aims to create intelligent machines that work and react like humans. So, (AI) is a technology that appears to impact (influence) human preference by learning, understanding complex contents, enhancing humans in executing both routine and non-routine tasks. In the future, (AI) technology that can be virtual personal assistant, as well as it may exist, such as robots with human-like processing capabilities.

How can (AI) technology impact global economy growth over the next 10 years? During this time period, (AI) technology is predicted to have wide-ranging applications including: Machine learning that automates analytical model building by using algorithms that allow machines to operate without human assistance.

In global education aspect, potential applications include predicting cause-and-effect relationships from biological data, identifying new drugs, self-driving cars, and protecting against fraud, improved natural language processing that allows computers to continue to better analysis, understand and generate language to interface with humans using natural human languages. For example, transcribing notes dictated by physicians, automatically drafting articles and translating text and speech. So (AI) technology can be applied to education aspect to improve humans' knowledge level.

In visual art aspect, (AI) machine vision that allows computers to identify objects, scenes and activities in images. Current applications of (AI) machine vision include providing objective

descriptions for the blind seeing(visual) needs.
We except the economic effects of (AI) technology to include both direct GDP growth from sectors that develop or manufacture. (AI) technology and indirect GDP growth through increased productivity in existing sectors that employ some form of (AI). If (AI) technology is an increasingly critical component of more products, it will become an integral part of many people's lives. Thus, (AI)'s ability to influence economic activity, rather than the economic or development status of the region. (AI) has the potential to impact income classes and to bring significant gains to both developed and developing countries. For example, (AI) has the potential to optimize good production around the world by analyzing agricultural regions and identifying what is necessary to improve crop yields.
In estimating the future economic effects by (AI) technology innovation, it is important to note that it is challenging to accurately predict which applications of (AI) will ultimately be commercially successful. In micro level economic influence, we need to apply methodologies to estimate the economic effects of investment in firms developing (AI) technology since investment levels in a technology are a telling sign of the future potential of that (AI) technology.

- How can (AI) influence GDP of high income countries in the next ten years?

How (AI)'s development may affect the global economy over the next ten years. In fact, (AI) technology has the potential to affect business across the global in a wide range of industries in ways only a number of technologies have done in the parts. For example, (AI) technology's expected to be a useful tool for enhancing human capabilities and in some instances replacing functions, such as driving a car, adoption of broadband internet, mobile telephone, industrial robotic automation have served to enhance human capabilities.
However, significant public debate has focused on projections of (AI) technology's effect on the labor force. However, large companies prefer to invest in (AI) technological industry. For example, face

book's (AI) research lab., google machine intelligence lab. and micro soft machine learning and artificial intelligence research division are all making advances in (AI) technology and investing in the industry's top talent. Additionally, between 2010 year and 2015 year, nearly $5 billion in venture capital funding invested in firms across the global developing and employing (AI) technology (Facebook (AI) Research).

- How can artificial intelligence impact on workplace?

Modern information technologies and the labor economy growth of machines is powered by artificial intelligence have already strongly influenced the world of work in the 21 ST century. Computers, algorithms and software simplify every tasks and it is impossible to image how most of our life could be managed without them. How can be the information economy characterized by exponential growth replaces the most production industry based on economy of scales? What will the future world of work look like and how long will it take to get? Will the future world of work be a world where humans spend less time earning their livelihood? Alternatively, are mass unemployment, mass poverty and social distortions also possible scenario for the future, where robots, artificial intelligence systems play an increasingly central role? These questions concern how artificial intelligence further development . Can influence labor economy growth on workplace ? When the labor market has widespread impact on intelligence property, information technology, product liability, competition and labor and employment laws.

How (AI) technology impacts on labor workplace.

The future influence any organizations how labor economies use of (AI) can be analyzed, such as deep machine learning is based on a set of model high level data. Unlike human workers, the machines are connected the whole time in workplace. If one machine makes a mistake, all autonomous systems will keep this in mind and will avoid the same mistake the next time.

Over the long run intelligent machines will win against every human expert. Production robots have been replacing employees because of the (AI) technology. They work more precisely than humans and cost loss. Creative solutions like 3D printers and the self learning ability of these production robots will replace human workers, the automatic data recording and data processing, traditional back office activities are no longer in demand. Autonomous software will collect necessary information and will send it to the employee who needs it. Additionally, dematerialization leads to the phenomenon that traditional physical products are becoming software. For example, CD or DVDs are being replaced by streaming services. The replacement of traditional event ticket, e-travel ticket service products or hard cash will be the next step, due to the possibility of payment by smartphone. So, (AI) technology will impact human's daily life consumption behaviors in the future. For another example, transportation tools, such as boats and ferries and private vehicles will use sensors and navigating without human input. Taxi and truck drivers will become obsolete, the stock store applies to stock managers and postal carriers of the delivery is distributed by (AI) machine delivery method.

What is the relationship between (AI) and (CRM)?

- Can (AI) technology impact on customer relationship management (CRM) ?

Nowadays , (AI) is a technology almost as old as the computer industry itself, it is similar with the advent of personal assistants function to businesses and personal promotion channel, such as (Amazon's Alexa, Apple's Siri, Google's Assistant) image recognition (face book), personalized recommendations (Netflix , Amazon). Those innovations have been driven by a increase in processing power, lower cost hardware, and the exploding creation and availability of data. It seems, (AI) technology can impact global customer service management method.

How to forecast economic impact modeling to (AI) will affect global economy? Can human forecast business revenue growth and job creation (or destruction) based on (AI) applied to customer relationship management (CRM) activities? In addition to the economic impact on (AI) or (CRM) which can include an estimate of the economic impact attributable to sales forces customer base. What can economic benefits be brought to (CRM) from (AI) technology?

Artificial intelligence(AI) comprises a set of technologies that use natural language processing, machine learning, knowledge graphs, and other tools to answer questions, discover insights and provide recommendations. Computer systems can use (AI) hypothesize and formulate possible answers based on available evidence can be trained through the ingestion of vast amounts of content, and automatically adapt and learn from (AI) self mistakes and failures.

So, any business organizations (customer service departments) can provide efficient and effective customer relationship management of excellent customer service quality if which applied (AI) technology system. The different type of (AI) systems include: (AI) system platforms, machine learning (AI) based data preparation and enrichment tools, machine vision/image recognition, voice speech recognition, text analysis and natural language processing, bots , e.g. face book website and virtual digital assistance solutions, social media pattern analysis , sentiment analysis, advanced numerical analysis (e.g. IOT streaming , machine logs), supporting technologies, knowledge base dialog management, Q&A processing etc. different (AI) technology system customer relationship management (CRM) tools.

(AI) (CRM) of activity can include these categories, such as: corporate marketing, marketing operation, field marketing, customer support, digital commerce, customer analytics, customer influenced product or service design, product or service pricing, finance information, presentation, customer billing, inventory , logistics and fulfilment support, partner management etc. different CRM tools.

(AI) technology of CRM has been carrying on plan different stages to achieve CRM personal assistant tool for businesses. The stages are such as, in the beginning stage of (AI) projects in place, implement now, pilot phase next year in the final stage of (AI) customer relationship management tools are foreseeable future. So, this CRM technology has been improved to plan in different stages every year to prepare to achieve full capacity of CRM service quality for businesses to use in the future.

Hence, how to develop an estimate prediction of the economic impact (AI) technologies could have CRM activities, which depends on gathering macroeconomic information on business revenue and the basic marketing of business revenue and the basic markup of business expenses by major functions (customer support, marketing and sales , production etc.)

An economic impact model that can gather data together and forecast the results how (AI) artificial intelligence technology brings (CRM) customer relationship management benefits to businesses, e.g. surveys investigation includes IT spending by sample countries, GDP and population estimates and forecasts, revenue per employee and ratios of IT spend to GDP. Surveys (questionnaire questions) of forecast results are influenced by (AI) impact can include: results are projected from surveys and rely on estimates are made by respondents on the expected financial improvements in categories of (AI) –assisted customer relationship management activities. The forecast assumes that these estimates are correct; financial estimates are based on estimates of "first year" improvement from full (AI) implementation; forecasts are from planning to implement any artificial intelligence of customer relationship management (CRM) projects, the improvement forecast is of categories of activity , e.g. corporate marketing , digital commerce, and customer analytics. They are not estimates of ROI for the (AI) software. They rely on conservative estimates to which each of these entities might affect company revenue, expenses or productivity. They also rely on estimates of the penetration of software in customer relationship management activities . Net new jobs created are based on the ratio of new revenue to jobs required to support that revenue . They can assume that 50% of the net new revenue will support increases in labor and the rest will go for capital and other operating expenses that may replace jobs lost to automation.

In the future, some of the ways in micro economic benefits to any organizations. (AI) technology is expected to impact CRM activities include: Spending up sales cycles, improving lead generation and qualification solving customer support problems faster (raising service quality), helping companies improve brand campaigns and recognition, lowering costs of support calls when increasing resolution rates, lowering the cost of recruiting employees and partners, increasing revenue from optimized product marketing, optimizing price, distribution logistics and preventing loss through fraud detection. So, micro economic benefits view point, it seems that (AI) CRM technology can raise any companies economic benefits for care term.

Artificial intelligence enables machines or the in-build software to behave like human beings which allows these decisions and act. The advent of (AI) is leading , talking, making decisions and act. The advent of (AI) is leading to new technologies advances and transforming the economic and employment opportunities for humans in a positive way. (AI) related technologies can facilitate our live. For example, industrial robotics, robotic medical assistants, smart games, financial forecasting software, big data analysis, algorithms in health and bioinformatics, pilotless cargo places, drone ambulances and general purpose and workplace robots and others. (Disruptors technologies: Advances that will transform life, business and the global economy).

Artificial intelligence also known as computational intelligence is defined as " the human –like intelligence exhibited by machines or software. It is theorized that intelligence of humans can be described and intelligence machines or software can simulate it. These machines software can be reasonable , learn, perceive and process information, like human mind and thus facilitate human life. They can think and act for us. So, artificial intelligence is an interdisciplinary field of study including computer science, neuroscience, psychology, linguistics and philosophy.

However, (AI) research and developments have economically impacted many industries, such as robotics, telecommunications,

computer applications , health, finance, heavy manufacturing, transportation, aviation, e-service and e-commerce, military , music and movie, toys and games entertainment etc. industries.
In fact, many ideas, systems and technologies have been developing in the world of (AI) technology. However, which are net called or considered (AI) products, rather which are mentioned with their specific names, such as smart graphics, machine learning, e-commerce etc. (i.e. this is called (AI) effect).

What is influence to (AI) job nature change to change developed and countries countries digital economy

- How can (AI) technology influence digital economy?

Nowadays, (AI) related industrial applications will replace most human power in fields, including call centers, customer services and air cargo transportation. (AI) technologies also help weather forecasting based on repeated rainfall pattern (data) recognition, through robotics (i.e. floor cleaning, moving lawns etc.) transporting people and products with unmanned vehicles, sending space unmanned smart shuttles, developing robotic arms, predicting market values in stock exchanges by internet, making homes safer, helping elderly and disabled using robotic servants etc.

Among the (AI) related technologies , there are a few that significance for the impact on society and especially on digital economy . (AI) is particularly influential in machine learning. Such as robotics, transportation, finance, health and bioinformatics, e-commerce , e-games, big online data gathering and internet-of-things. For example, machine e-learning is based in bioinformatics and robots that can learn new skills for better caregiving in healthcare. What is machine e-learning? Machines can e-learn from e-data gathering, coming up generalizations and making decisions to act in certain ways from internet.
There are important applications , such as e-machine perception, electronic online natural language learning processing, online search engines, online bioinformatics, online brain –computer

interface, online game playing, online robot locomotion, online advertising, online computations finances, online health monitoring, online DNA classification and decision making, online in chemistry –cheminformatics . So, online machine learning can positively impact productivity and it can enhance information and analytical system from (AI) online channel.

What is robotics? Robotics is one of the most strongly influenced fields in (AI). For example, heavy manufacturing industries, robots and used and man power is replaced for effectiveness, precision, and accuracy, especially in respective or dangerous tasks, including welding, assembling , picking and placing .

So, robots can acquire new skills or adapt the changing dynamic environment. Also, artificial intelligence can be applied in developing transportation. For example, automated vehicles, driver assistance systems , safety systems, collision avoidance systems and public transportation. Moreover, (AI) technology has proven to produce some of the best tools to predict stock market fluctuations from internet data gathering method. It's predictions are based on ever-evolving predictions algorithms and systems learn new models and make connections between historical data and new data to measure stock market trading more accurate from internet data gathering channel.

In health field, especially in health data processing , analysis, decision making support and medical diagnosis. So, online data can show which patients will need what treatment and what alternative drugs could be used more accurate from (AI) online data gathering method. Bioinformatics is an interdisciplinary field combining statistics, (AI) online technology can help in discovering data patterns and modeling through the application of machine learning, artificial neural networks and genetic algorithms. For example, further (AI) technology development of human genome project of online data sequences.

Online shopping can be facilitated by virtual assistants developed through (AI) technology and these assistants can offer the best advice. (AI) online purchase coming after every product image recommendations and personalization bring important revenue to shopping online sites, like Amazon . Smart computer graphics and games, artificial intelligence is useful in smarter computer, graphics, scene modeling , scene rendering processes in order to create, for example, effective human –robot interactions , online machine learning, online strategic games techniques etc. online computer related (AI) software.

So, online big data analysis and big data does have a critical need in the world of online intelligence machines and software in our future. In other words, (AI) offers online technology to enable online big data analysis to provide industrial organizations with valuable information for effective decision making in short time. For example, what IBM's Watson achieved: this machine used 200 million of structured and unstructured content with a special technology of hypothesis generation, massive evidence gathering, analysis and scoring from internet channel.

Finally, (AI) online technology another related internet invention (internet of things) (IOT) is the network of machines or objects connected through internet. These connected objects can sense their internal and external environment, communicate with each other, can send critical data and finally can make decisions to act or correct their environment from (AI) online technology. For example, factories can monitor and automatically change production processes, hospitals can monitor and regulate the health conditions of their patients , schools can collect data from facilities and cars can send data to car makers from (AI) online technology.

Partner predicts that (IOT) market will create about trillion amount value by 2020 year. Although machines collect big data from their environment, whether which gain an insight or learn from these online data largely depends on the (AI) online machine learning principals and (AI) online technology. In 2013, Mckinsey estimated that disruptive technologies closely related with potential economic impact in 2025 year between \$7.1 to \$13.1 trillion amount (automation of knowledge work, advanced robotics, autonomous or near-autonomous vehicles).

What is the relationship between
(AI) and global digital economy development ?

● Could work activities in China be automated
making in the nation with the world's largest automation potential? Can (AI) technology influence China economy? Could China workers be affected and jobs made up of routine work activities and predictable? Will programmable tasks be particularly impact to China employment market ? When impact on labor market is likely to be gradual at the aggregate level, it can be sudden and dramatic at the level of specific work activities, rending some job obsolete fairly. Overall (AI) technology will raise digital skills when reducing demand for medium incomer inequality for China workers. It seems (AI) technology's effect on productivity could be crucial to China's future economic growth as the population ages are increasing.

In China, some biggest technological companies driving significant investments in research and development. Moreover, China is one of the leading global (AI) technology development county. However, China will need to focus on building its innovation capacity. For example, United States and United Kingdom are currently producing more influential (AI) technological research. However, if China planed to achieve (AI) technology success, it's traditional industries will need to develop technical know-how –to and overcoming implementation costs prepare to develop (AI) . When (AI) technology is introduced into China society, China government needs to raise concerning ethical, legal, technological security etc. business questions. Also, surrounding issues include privacy, discrimination, legal liability and regulation. It aims to encourage overseas investors to choose to invest (AI) technological industry to raise GDP growth and manufacturing industries income growth for long term in China.

If China encouraged overseas (AI) technology investment in its country. It is possible to influence China employment market to be changed. Because (AI) technology will impact to influence China people daily life. Due to (AI) technology is introduced to China society, many rich people will prefer to spend to buy any high (AI) technological products for entertainment or learning or machine man driving etc. daily necessity activities. Then it will raise GDP growth and will raise (AI) manufacturers or related-(AI) technological manufacturers profit. It is beneficial to China because it can become one high knowledgeable and (AI) technological economical society.

But it will bring bad influences to raise unemployment chance for the low skillful labor. In labor economy aspect influence , how (AI) technology can influence China low skillful labor unemployment ratio raising. The raising low skill labor unemployment reason is because China low skillful human labors are argued or are replaced by (AI) technology creating new challenges to introduce to influence China society of simply human manufacturing job nature to be changed to be high (AI) technology manufacturing job nature in any China factories. Moreover, when (AI) technology introduction to China, it will cause other related social challenges in China. The varied (AI) related challenges, including the difficulty of creating safe and reliable hardware for sensing and affecting (transportation and education), the challenges of gaining public trust, a low resource comities and public safety and security, the challenges of overcoming fears or marginalizing humans in China employment and workplace and the risk of diminishing interpersonal trust because the low skillful labors won't believe any China employers will give chance to employ them , due to (AI) technology will replace their skills and man manufacturing of productivity is much less to compare to (AI) technology manufacturing method.

- How does (AI) technology influence
the future of employment change?

Are future nature of jobs changed to computerization from (AI) technology? Where are the probability of computing occupations from (AI) technology influence? What is expected impacts of future computing on labor market from (AI) technology influence? John

Maynard Keynes's frequently cited prediction of widespread technological unemployment " du to our discovery of means of economic the use of labor outrunning the pace of which we can find new used of labor" (Keynes, 1933, p.3).

In the future, (AI) technology will impact some nature of occupations to change computing. This chance will also influence some countries' economic change. For example, some factory human labors hand routine manufacturing tasks will be changed to computerization of routine manufacturing tasks by (AI) technological machine men hand manufacturing method. it will cause a structured shift in the labor market, with workers reallocating their labor supply from middle-income manufacturing to low-income service occupations.

Arguably, this is because the manual tasks of service occupations are less computerization, as who require a higher degree of flexibility and physical adaptability. So, (AI) technology will influence the human hand labor skillful occupation nature of task cheaper , such as vehicle manufacturing , ship manufacturing, computer manufacturing, steel manufacturing, television, radio etc. home electronic products of heavy machine industry change. Due to (AI) technology machine man will be proper to be used to manufacturing these electronic products when the (AI) technology innovation can develop to the mature stage. Then, any countries manufacturers will choose to use (AI) technology machine man, instead of human hand production.

Supposing the future prices of computing are fallen, seriously, problem solving skills are becoming relatively productive, explaining the substantial employment growth in manufacturing occupations, involving cognitive tasks where skilled labor has a comparative advantage, as well as the increase education needs for (AI) technology computing of machine man subject study.

Prediction of education needs for (AI) technology student numbers will increase, due to manufacturing industry needs many (AI) technology students in future employment market. Another (AI) technology influence if the future (AI) technological innovation, e.g.

machine man manufacturing or machine man service industries will both increase demand, then with more sophistic software technologies will be disrupted labor markets by marketing workers redundant.

For publishing industry, what is striking about the case in paper book publishing industry will be unpopular? Due to the electronic book publishing industry will be popular, e.g. Amazon publish . (AI) technology can influence paper book manufacturing method which is replaced by machine man electronic book manufacturing method as well as it will cause the computerization is no longer confined to routine manufacturing tasks. Due to (AI) machine man manufacturing technology will be proper to be used to manufacture any products in short time efficiently and effectively , e.g. electronic book products. In the future, if it is fact to occur this case, such as (AI) technological machine man manufacturing method will be adopted (applied) to manufacture electronic books or any products in possible. (AI) technology will cause many manufacturing workers are unemployed. It is beneficial to employers, who can reduce to spend much wages expenditure to employ manufacturing workers, but it will cause many manufacturing workers loss jobs and reduce income to support whose families lives. It will cause social challenges, e.g. increasing stealing crimes if the manufacturing workers had not other skills to find other jobs to do easily. So, manufacturers need to concern over technological unemployment which will be hardly future phenomenon if who decided to dismiss all manufacturing workers, due to (AI) technology machine men replace to them.

If (AI) technology can be innovated to produce any kinds of machine man to serve any service or manufacturing industries successfully. Then, it will bring these questions: Can future that workers be influenced to be automation employment and productivity by (AI) technology influence? Does it impact to influence the (AI) technology countries' productivity and growth and natural resources development and labor markets and evolution of global financial markets and economic impact of technology and innovation and urbanization etc. issues? How will automation transform the workplace? What will be the implication for employment? What is likely to be its impact both on productivity in the global economy and on employment?

In fact, automatic of activities can enable businesses to improve performance by reducing errors chance and improving quality and speed, and same cases achieving outcomes that go beyond human capabilities. Some economists indicate (AI) technology would give a needed boost to economic growth and prosperity have of the working age population in many countries. Based on the scenario modeling, they estimate automation could raise productivity growth globally by 0.8 to 1.4 % annually. They also indicated that almost half the activities people are almost $1.6 trillion in wages to do in the global economy have the potential to be automated adapting current demonstrates technology, according to their analysis of more than 2,000 work activities across 800 occupations. When less than 5% of all occupations can be automated entirely using demonstrated technology, about 60% of all occupations have at least 30% of worker made activities, that would be automated. More occupation will change to be automated. They also indicated for business performance benefits of automation are relatively clear, but the issues are more complicated by policy making to attract foreign investors. Beyond technical feasibility, the cost of technology, competition labor will include skills and supply and demand dynamics, performance benefits and beyond labor cost savings and social and regulatory acceptance will affect the automation. Their predictions suggest that half of today work activities could be automated by 2055 year, but this could happen 10 to 20 years earlier or latter depending on the various factors in addition to their wider economic condition.

Some scientists suggest (AI) technology is finally starting to deliver real-life business benefits. Computer power is growing significantly , algorithms are becoming more sophisticated and perhaps most important of all, the world is generating vast quantities of the fuel that powers (AI) technology data billions of gigabytes of it every day.

Also, online firms are digital natives, such as Google online search service company is investing on (AI) technology. For new though most of the news if coming from the suppliers of (AI) technologies. And many new users are only in the experimental phase. Few products are on the market or are likely to arrive these soon to drive immediate and widespread adoption. As a result, analysts believe (AI) technology's potential will give true economic benefit in the future. (AI) industry will introduce to suppliers and users to raise economic potential of (AI) technology.

In the future, (AI) technology systems can solve business problems. Some scientists categorized those into five technology systems that are key areas of (AI) technology development: robotics and autonomous vehicles, computer vision language virtual agents and machine learning , which is based on algorithms that learn from data without replying on rules-based programming in order to draw conclusions or direct an action.

Such as computer vision and language includes natural language processing, analytics, speech recognition technology, some are about learning from information, such as about machine learning and others are related to acting on information, such as robotics, autonomous vehicles and virtual agents, which are computer programs that can converse with humans. Machine learning and a subfield called deep learning are artificial intelligence applications.

- Can artificial intelligence impact

global economy growth?

Artificial intelligence (AI) is a term first defined in 1956 year. It is a branch of computer science that aims to create intelligent machines that work and react like humans. In contrast today, 60 years later, (AI) is characterized by a number of applications, including computers playing games against humans and understanding human languages, virtual personal assistants, and robotics which involve computers seeing , hearing and reacting to sensory stimuli. In the future, technologists predict for (AI) technology ranging from (AI) being used as a tool to aid relatively simple processes for robots with human like mental capabilities, who expect (AI) technology can emulate human performance by learning, coming to mind its own conclusions, understanding complex content, engaging in dialog with people, enhancing human cognitive performance or replacing humans in executing both routine and non-routine tasks. In existing industry, (AI) technology is used , such as targeted advertising and virtual used personal assistant as well as the (AI) technology that my exist in the future, such as robots with human vehicle processing capabilities.

The range of (AI) technology's progress in the future will determine the economic impact future of (AI) technology on the global economy with more limited advances and applications (i.e. weak (AI) only) corresponding to more limited economic impacts and more substantial progress, i.e. strong (AI) technology is corresponding to more significant economic impact.

(AI) technology learning that automates analytical model, including predicting cause-and-effect relationship from biological data, identifying new drugs, self-driving cars and protecting against fraud etc. functions. Also (AI) learning can improve natural language processing that allows computers to continue to better analyze, understand and generate language to interface with human using the natural human language, virtual personal assistant, helps users by providing scheduling appointment, reminds organizing personal finance and finding providers of various services, machine vision allows (AI) machine man to identify object, scenes and activities in detect pedestrians and bicyclists.

We expect the economic effects of (AI) technology to include both direct GDP growth from sectors that develop or manufacture (AI) technology and indirect GDP growth through increased productivity in existing sectors that employ some from of (AI) technology. If (AI) producing sectors could grow, then it could lead to increase revenues and employment of (AI) technological professionals within these existing firms as well as the potential creation of entirely new economic activities to any countries' societies productivity improvement in existing sectors could be realized through faster and move efficient processes and decision making as well as increased (AI) technological knowledge and access to information available in societies easily.

In the future, if (AI) technology is an increasingly critical component of more products, it will become an integral part of necessary products of many people's lives. The extent of (AI)'s economy effort is also likely to vary from region to region, thought variation may be more dependent on the predominate economic activity of a region and the (AI) ability can influence economic activity, rather then the economic or developmental status of the regions. (AI) technology can move accessibility and can use source development to do international business between one country and another country.

So (AI) technology has the potential to give benefits to different income chooses and to bring significant gains to both developed and developing countries. For agricultural technology, (AI) has the potential to optimize food production around the world by analyzing agricultural regions and identifying what is necessary to improve crop yield. In total, (AI) technology gives greater economic impact to any countries agricultural regions if which implemented (AI) technology to grow crop , fruit etc. food production in the farms. Investment in (AI) technology is such as capital investment to any countries' public or private enterprises. So, it will have large economic impact to the future . If the (AI) technology is reasonable invested to the different needs aspect by the public or private enterprises in the country. Then, it will have good economic impact to the country in the future. However, when (AI) technology is likely to affect both the productivity and employment components of economic growth in many sectors. Significant public debate has focused on projections of (AI)'s effect on the labor force. However, for instance, some researchers have argued that the rise of (AI) technology and automation will led to significant unemployment as capital is substituted for the low skillful labor. So, they point to the concern that the increasing sophistication of (AI) technology may balance skilled and semi-skilled workers and the reduce the size of the middle class. However, this is not a new argument, due to (AI) technology negatively affecting the labor force and leading to mass unemployment. Because the (AI) technology is the substitution of machinery for human labor. Although, employment in certain industries, has been reduced in the past due to technological advancement. For long term, the labor market has adapted to the introduction of new technology, giving rise to new jobs in new areas. (AI) technology may also be accomplished without a reduction to total employment in the long-term to some Asia countries, such as Hong Kong and Japan. Because Hong Kong and Japan many low skilled labor, e.g. security, cleaner who complaint that employers need them to work long time hours. (abnormal working hours) e.g. one day 12 to 15 working hour per day. Hence, if (AI) machine means invention technology success. Security or cleaning job can be worked from (AI) machine man in some hours every day in order to reduce the long time working hours cleaners or security workers, e.g. one (AI) machine man works 4 hours for cleaning or security job, one day as well as another cleaner or security labor only needs to work 8 hours one day. So total security or cleaning employers can employ 12 hours machine cleaners or security workers and human cleaners or security workers in one day. For long term benefit, Hong Kong or Japan every security or cleaning worker does not need to work 12 hours minimum working hours one day. They won't feel tried and bore and without private with whose families, so who will accept to do these cleaning or security jobs, even they can raise work efficient and performance when who feel happy and health.

So, (AI) technology of machine man invention can raise low skillful labor efficiency and it can help them to avoid abnormal working hours demand in some busy work life countries, such as Hong Kong and Japan. Before, one Japan female labor feel unhappy to work, due to who often needs to work abnormal working hours for her employer and who has less sleeping and without any private time to enjoy her life with her families every day. So this abnormal working hours factor causes her to do commit suicide behavior, then she is die unlucky. So (AI) technology of machine man invention ought avoid abnormal working hours demand for employer in any countries in the future.

The most important occurrence to any employers, some researchers had attempted to do one experiment to find that private research and development , venture capital and public research and development investment all have strong net effect or economic growth with venture capital funding further having the strongest such effect from (AI) technology. The researchers hypothesize the venture capital investment contributes to economic growth through (AI) technology innovation and by the capacity of an economy to use existing (AI) technology knowledge to increase productivity. They predict the impacts of venture capital, business-research and development and public research and development can raise multi factor productivity from (AI) technology introduction.

Can (AI) technology influence the economic development to developing countries? The developing regions of the world contain most of natural resources. If one day, (AI) technology has invent one kind of machine man which can assist any gas or oil workers to seek any new oil/gas natural resource locations easily. I believe that (AI) technology can help these natural resource exploitation countries will gain economic benefit more easily. So, (AI) driven technology can be used to change to create any new opportunities to address poor management or resources and improve human well being, such as Africa Latin America and India can use (AI) technology machine man to seek any oil/gas natural resource countries exploitation activities to attempt to gain much economic benefits.

- Why will (AI) technology grow economic development ?

Nowadays, increases in capital and labor are no longer driving the levels of economic growth, such as (AI) technology. The ability of increase in capital investment and in labor of traditional drivers of production, have no longer to be enjoyed in most developed economies ,e.g. developed country, US, UK . However, artificial intelligence has the potential to overcome the physical limitation of capital and labor to avoid missing out on this opportunity. So, policy makers and business leaders must prepare for and work toward a future with artificial intelligence. They must do with the idea that (AI) is another simply method to enhance productivity method . Rather they must see (AI) as the tool that can transform thinking about how growth is created.

Economists have always thought of new technologies are as driving growth their ability to enhancing. It can replace labor and capital factor of production. So, it brings this question: What is the factor of production (AI) technology characteristics. They key factor is to see (AI) technology as a capital-labor .

(AI) can replicate labor activities at much greater scale and speed, and to even perform some tasks began the capabilities of human. For example, by using virtual assistants , 1000 legal documents can be reviewed in a matter of days instead of taking three people six moths to complete. Some (AI) technology may be one kind of factor of production in the future. For another example, people will work in workplace digitalization environment. So, in the future, working environment and information management are automated. Such as Konica camera sale company will use workplace digitalization. So , (AI) technology can provide workplace digitalization in order to raise productivity efficiency. (AI) technology will be one kind of production which is replaced by workplace digitalization and it will grow any organization productivity efficiently. Then, (AI) technology will assist overall social economy growth , due to productivity is raised and products can be produced in short time to prepare to sell in consumption market. So, time will be shortened to increase GDP growth fast for the development of (AI) technology countries.

- How can (AI) technology impact to global economic and social and psychological changes?

What will be the development of (AI) technology and predictions concerning the future evolution? The computers and robots will develop conscious, intelligent and minds into humans, enhancing psychological and behavioral abilities and allowing for direct communication with (AI) minds. (AI) technology will be impacted human life by (AI) technology information communicative and environmental influence. A " world brain" and " world mind", this psychological system will be enhanced and enriched the capacities of both individual and collective cognition by (AI) technology of service industries.

(AI) technology with influence these human needs of service industries changes, such as , biological science, finance, entertainment, business, biological science, transportation, communication military etc. The personal computer evolution, the internet and the world wide web which exploded on the scene, linking business, homes, schools, social organizations which were a completely unpredicted phenomenon to influence human life. Kurzweil (1999) predicts that by 2029 year, most human communication will be with machines. According to Person, by 2100 year, there will be human machine convergence.

How can (AI) technology influence environmental protection to make benefits to farming economic growth? (AI) technology can be applied to predict how to solve environmental pollution challenge to avoid to damage any crop or vegetable or rice or fruit etc. food growth. Because environmental experts can gather global environmental pollution data from an environmental database to build a perform a systematic analysis from (AI) technology. The first step is this broad analysis can include understanding, statistical and data gathering techniques to obtain the relevant data, the correlation among the variables involved, and a list of possible models. The next step is to select a set of methods and models that cover all kinds of knowledge and functionalities needed for the decision making process. Once the models are selected, they must be fully implemented by means of machine learning , data mining, statistical or numerical technique. After that, those models must be integrated to build the whole EDSS. The EDSS must be tested to check its performance, accuracy, usefulness and reliability, both from the user's and (AI) technology/computer scientist's point of view. If these is any wrong feature in any development stage, such as model's integration, models' implementation, selection of models, database, problem analysis etc. the developers must come back in the update th required components. When the evaluation phase is all right, the EDSS is ready to be applied to the environment. The great contribution of artificial intelligence to EDSS the integration of several methods complementing the classical statistical models/ simulation , statistical analysis, linear models, etc. and numerical models (control algorithms, optimization techniques etc.) .

This cooperation makes the resulting systems more reliable and powerful in coping with real world environment systems. Date interpretation has been a principal area of research in (AI) technology since the very beginning. The most demanding problem in the environmental assessment context. Knowledge representation permits the definition of the different types of data that the existing methods adapt to the process. There is also a lot of work to clean, repair and transform the huge available quantities of raw data. Apart from this, the availability of meta-information or background knowledge is required to guide the process. Data mining is multi-disciplinary: It covers expert systems, data based technology, statistics, data visualization and unsupervised machine learning. These techniques operate at the level of data and background information, where numerous and often incompatible new commensurate pieces of information from disparate sources have to be brought together (K, Fedra, 1994).

So, it seems that in the future, (AI) technology with the increasing maturity in particular those related to knowledge and engineering, new dimensions can be assisted to users in environmental decision making are available. For example, many environmental systems are characterized both by incomplete models and by limited data. Hence, in the future, (AI) technology will be applied to predict climate change to reduce crop or fruit etc. food agriculture challenge by climate change bad influence.

● Will (AI) technology influence digital economy change to manufacturing industry ?

To understand how the manufacturing business must adapt to prosper in the technology, we need to understand how (AI) technology will change us to shape our daily habits to satisfy our expectation of products to how we shop and even the immediate of the entire process. For example, taxi services are in the crosshairs as on demand transportation services like, available of the touch of a smart phone button expand. In fact, Yellow lab, US country , san Francisco city's largest taxi company is filing for bankruptcy as the industry starts to change faster than almost anyone expected. However, at this point, its more than an app that is changing, some our taxi passengers renting taxi transportation to catch consumption behavior.

(AI) technology will influence digital economy for taxi passenger's individual customer experience, offering a growing renting taxi to catch of service and feedback opportunities when any one taxi passenger who chooses to use mobile phone app online tool to prepaid to rent any taxi more easily.

Also in the long term, (AI) technology can influence vehicles drive themselves of behavior. Already, companies like Google and GM are working on projects to bring fleets of autonomous vehicles to cities at the path of a button.

Moreover, this on-demand service model is beginning to appear across a much broader range of markets. For example , Amazon company is investing in its own fleet of trucks, planes and even drone at the same time as it pushes for same-day delivery of products. As some point, vehicles will be autonomous too. So, it seems that (AI) technique will influence any transportations choose to use digital autonomous driving technology in the future . For Amazon company case, it is not stopping of logistics. It is also aiming to automatically manage the supply of consumer home products with its recently launched Amazon replenishment service, Dash. Dash is a digital service that enables that connected derive to automatically order physical products from Amazon when supplies are running low. So, it seems (AI) technology will be applied to logistic function by digital technology method introduction in the future.

Hence autonomous vehicles will optimize industry supply chains and logistics operations through increased efficiency and flexibility. In fact, fully automated and lean supply chains will keep reduce load sizes and inventory by leveraging smart distribution technologies and smaller autonomous vehicles by machine man assistance. If Amazon continues to grow market share for online sales by reducing effort required by the consumer to place an order, when also contributing the almost immediate delivery of products to the doorstep. So, it will further fuel the trend toward on-demand derive. As Amazon company fuels the on-demand economy, consumers will expect immediacy in more parts of the digital economy. On top of speed, consumers increasing expect more personalization options.

So, (AI) technology will influence digital manufacturing, such as Amazon publishing to monitor every aspect of every process in real -time and communicating to self-optimized deep learning robotics, new methods of high volume and high customization will become possible. Then, as products merge into product platforms and even services, manufacturers have the opportunity to provide components and platforms used by smaller players. So, (AI) technology will influence manufacturing industry to choose automated SMI lines, robots installed, automation engineers.

Another future (AI) technology development can be applied to space science aspect, such as Automation engineering space in manufacturing process to achieve digital manufacturing benefits to any businesses in the future. Such as reducing cost, shortening manufacturing time, raising efficiency, shortening delivery products to client individual time. How can artificial intelligence give the need and advanced fast and evaluation methods benefits for space exploration? When US NASA (space exploration organization) achieves any space exploration missions, it will

answer this question:
When is it useful to have a machine use (AI) technology to achieve a decision? After all, after millions of years of space exploration and rough 10,000 years of civilization, humans are usually quite good at making decisions in complex uncertain environments. Through, Johns Hoplains University's Applied Physical Lab. Research in (AI) technology enabled systems, which has identified three general use cases for (AI) technology to explore space mission:
First, for some tasks (AI) technology is more cost effectiveness than human. Second, (AI) technology is better suited than humans at solving some, but not all problems. Third, (AI) technology allows NASA organization's space exploration mission to develop machines that ate capable of responding faster than when a human is in the decision loop (D. Scheidt, 2012, A. Castano et. al. 2008).

So, the use of (AI) technology to enable science by observing the pace of rapidly evolving phenomena was demonstrated. It is more effectively coordinating and (AI) technology utilizing to earn economic benefits to use for space exploration mission.
However, (AI) technology also have current risk for space exploration. Today (AI) technology is immature and requires further development to reach its potential. For instance, the (AI) technology algorithms that detected the dust derive could not have identified whether the Martain weather represented a threat to the cover. Also it can not yet use instrument input to determine what, where and how to autonomously make the next space science measurement. An equally important factor limiting (AI)'s deployment is that lacks the methodology and technology to effectively test (AI) technology. So, the challenge will testing (AI) enabled system is how (AI) performance can be measured. It would be NASA organization's difficulty to find (AI) technology to develop to carry on researching any space exploration missions in the future. However, (AI) technology will be a good economic benefit choice for space exploration mission in the future.

- What is artificial intelligence potential
benefits and ethical considerations?

The ability of (AI) technology systems to transform vast amounts of complex information into insight has the potential to help solve manufacturing or service challenges for human needs. However, to reap the societal benefits of (AI) systems, humans will need to trust then and make sure that which follow the same ethical principles, moral values, professional codes and social norms that we humans would follow in the same scenario, research and educational efforts as well as carefully designed regulation in order to achieve the most effort of economic benefits goals. For example, international business machines corporation (IBM) is actively engaged both competitors , in global discussions about how to make (AI) ethical and as beneficial as possible for people as social economic benefits.

(AI) is usually defined as the " capability of a computer program to perform tasks or reasoning processes " that human usually associate to intelligence in a human being. Often, it has to do with the ability to make a good decision, even when there is uncertainty, too much information to handle. As an example, play chess or complex card games of entertainment activities is believed to need some form of intelligence in a human being, as well as choosing the best medical facilities in a difficult medical case, or creating something new, such as mathematical theorem or even some form of act, or even driving automatic machine man (self driving vehicle) replacing human driving in the middle of a crowded city.

(AI) needs depends on what we consider being intelligence in the behavior of a human being act a certain point in time. If human belief about human intelligence changes and we don't believe any longer that a certain task requires intelligence, then a computer program performing that task is no longer part of (AI), it becomes just another boring computer program. So, it means that (AI) technology will replace some old computer programs, if human can invent new generation of (AI) software for any functions or activities to satisfy human needs.

As IBM, it argues intelligence. This means that we aim to build systems that enhance and scale human expertise and skills rather than replacing them. We therefore focus on practical applications of

(AI) capabilities that assist people in performing well-defined tasks of needs by exploiting and wide range of (AI)-based services. We also use the term " cognitive computing" it is mean a comprehensive net of capabilities based on technology. It comprises the fields of machine learning, reasoning and decision technologies, language, speech and vision recognition and processing technologies, high performance and high efficient functions for any industries or individual consumers needs. For example, robotics, which are usually very good at doing what which are supposed to in any environment, much have public shopping center, factory etc. places which need simply services from the robot (machine man), such as cleans the floor of our houses to the robot that can work together with humans in production chains, passing through the warehouse, robots can take care of the tasks of an entire warehouse and the companion robots like Nao, Pepper, Aibo and Giraff, who can entertain use, talk to use and help elderly people to stay connected to their friends, relatives and doctors.

Google company is building automatic machine (self-driving cars) and has acquired more than 10 robotics companies. Facebook had opened whole new research facility only on (AI) research. Apply computer has developed Siri. Microsoft computer company has built a similar personalized assistant. Google has Deep mind, a UK company whose long term aim is to build general (AI) and has already great potential to win game to the world champion and IBM is investing a huge amount of resources in applying its Watson cognitive computing system to the medical domains to finance and to personalized education. In Europe, IBM is establishing new centers in Munich and Milan focused in the application of cognitive computer capabilities to the internet of things and healthcare respectively.

For example, automatic machine man (self-driving cars) are all about (AI), which used to be able to see what happens in the street (signals ,lanes, other cars, pedestrians, traffic lights, which need to able predict what other cars and pedestrians will do, and who need to be able to cope with unforeseen situations. Since, most car accidents are due to human fault, it is estimated that the adoption of self-driving cars will save about half of the lives that are usually last in car accidents.

IBM Watson company has to understand spoken language, make sense of massive amount to text , respond correctly to questions in many categories, as well as assess its own confidence in responding to such questions. In the future, (AI) technology can own question/ answering capabilities that would be very useful, for example, in assisting a doctor when trying to some to the correct diagnosis for a patient and to propose the best therapy .

Intelligent machines can also rely on huge amounts of data to be used to learn how to make better decisions. This data comes from all of us over the years Facebook users have uploaded more than 250 billion pictures and every day who upload about 350 million more. Every second, we submit 40,000 google search queries. So, (AI) technology will be connected through the web from appliances to traffic lights from cars to watches. Other tasks that are very easy for humans are physical and manipulation tasks, such as walking , running, picking up an object to make its shape and location, restricted environment. But (AI) machine man technology still not able to have the general physical and manipulation capabilities even of a 6 year old.

So, it brings this question: Why do (AI) scientists need to concern ethics? Because (AI) technology is complex, information into insight has the potential to reveal long held secrets and help solve some of the world's most difficult problems. (AI) systems can potentially be used to help discover insights to treat disease, predict the whether, and manage the global economy. So, ethic issues is important to and (AI) scientists . If any one new (AI) technology research investigation could success, it will be a secret to and the (AI) scientists can not permit to their loyalty to any competitors to damage the fair (AI) technology products trading market. The country (countries) (AI) technology scientists need to concern ethic issues, who need to keep secrets for their countries economic or/and social benefits. This is moral issues to any countries/country loyalty is whose countries intangible assets. They can not sell (AI) loyalty to any their countries to assist whose economic benefits immorally.

- How can (AI) technology influence to global health care economy development?

According to (AI) lecturer analysis, when combined key clinical health (AI) application can potentially create $150 billion in annual savings for the US healthcare economy by 2026 year. (AI) technology is re-winning modern conception of healthcare delivery. It enables machines to sense, comprehend, act and learn. So which can perform administrative and clinical healthcare functions (Accenture, 2017).

It will help health care service organizations to reduce health care cost, will improve and raise service quality and access. So, (AI) health market size will be predicted growth. (AI) applications in health care include robot-assisted surgery, virtual nursing assistant, administrative workflow assistant, fraud detection, error reduction connected machines, clinical trial participant identifier, preliminary diagnosis, automated image diagnosis and cybersecurity.

What kind of benefits (AI) technology can contribute to healthcare service? (AI) technology can deliver what many health care organizations need, such as financial and operational of labor costs, digital expectations from patient consumers how to use (AI) technology to solve interoperability challenges in any healthcare organizations. Also (AI) technology can be applied to wellness an d lifestyle management, diagnostics, delivers financially but also way of organizational and workflow improvement. So, (AI) technology will be continue to become most prevalent and adoption to healthcare organizations , which must need to enhance structure to be position to take full advantages of new (AI) technological capabilities. (AI) technology can change the nature of work and employment is rapidly changing to make the best use of both humans and (AI) talent in healthcare industry in the future. For example, (AI) technology offers a way to fill in gaps and the rising labor shortage in healthcare. According to Accenture analysis, the physicians shortage is increasing. However, (AI) technology will manufacture healthcare machine men to replace physicians in future one day(2017). Hence, (AI) technology will be invented to raise health care service staffs work efficiency and performance in any hospitals or clinics in the future.

In conclusion, (AI) technology will raise efficiency for any service or manufacturing industries in the future, although, it is possible that it will also rise low skillful workers unemployment numbers. But, the most important influence to human technological innovation will be risen and it will influence human life will be changed to be better, e.g. self drive cars, health care physician machine men, machine man cleaners etc. intelligent machine men will be manufactured to serve for our daily life. Furthermore, (AI) technological products will influence countries trading, some low technological development countries manufacturing businessmen can choose to buy any (AI) products to raise whose productivity and efficiency and reducing cost to achieve economic cost saving result. Also, GDP of trading growth income will increase to the (AI) products sale countries. Hence, it will be beneficial to economic development to both developed and developing countries both in the future as well as (AI) scientists time and money spending will be valued to continue to invest (AI) technology development for human life and economy benefits for long term.

In conclusion (AI) technology will raise macro economy growth and it can create many (AI) jobs , but it also raise the low level technological worker unemployment change. In the future, (AI) technology can be applied to digital technology to attempt to invent any new undiscovered (AI) and digital technology. So, it needs any scientists to continue to research how digital and (AI) technology can be mixed to satisfy human's future undiscovered needs.

Must Developed And Developing Countries Need Artificial Intelligent To Replace Human Job

Must developed and developing countries need artificial intelligent development? If one developed country, e.g. US, UK ,

Japan , Singapore it does not continue to develop artificial intelligence, robotic, then what disadvantges or weaknesses , it will encounter to compare when it chooses to continue to develop this artificial intelligent technology in society. If one developing country, e.g. China, Korea, Taiwan, it does not continue to develop artificial intelligence, robotic, then what disadantages or weaknesses, it will also encounter to compare when it chooses to continue to develop this artificial intelligent technology in in society. I shall explan the reasons why the results may cause to either the developed country, or the developing country as below:

- How AI help developing countries to communication and agriculture and learning and medical delivery development

Why can AI help developing countries ? Drones that pick inaccessible crops and mobile phones that give medical advice are two of the ways AI can transform life in the developing world. Artificial intelligence (AI) may improve the lives of the world's poor, the technology needed to revolutionise inefficient, ineffective food and healthcare systems in developing countries is well. For example, in low-income areas, agriculture and healthcare are two critical ecosystems that we can apply AI to immediately; this is not the far future, or even in five years.

Artificial intelligence (AI) has seeped into the daily lives of people in the developed world. From virtual assistants to recommendation engines, AI is in the news, our homes and offices. There is a lot of potential in terms of AI usage, especially in humanitarian areas. The impact could have a multiplier effect in developing countries, where resources are limited.

Emergency Response to developing countries' earthquake natural damage suddence occurrence predicting

AI and machine learning are still finding importance in emerging markets, but certain applications have emerged and are now widely used. For instance, predictive models for disaster relief enable first responders to automatically analyze large-scale behavior and movement through multiple sources of data including social media platforms, web forums, news sources, etc. Based on collected data, responders can scale reconstruction efforts and distribute supplies in a timely manner.

Why and how AI can assist farmers to predict when the earthquake occurs suddenly in order to avoid or reduce the natural damage to their agriculture productive number loss. For example, In 2015, when a

major earthquake hit Nepal, more than 8 million people were affected. During the aftermath, drones were used to map and assess the destruction and speed up the rescue mission. The town of Sankhu, situated about 20 kilometers northeast of Kathmandu, was among the highly affected locations. In May 2018, my company Fusemachines and GeoSpatial Systems partnered with Sankhu's city officials to use drones and artificial intelligence in an effort to automatically estimate the reconstruction need. After processing data accumulated from a drone-powered aerial mapping of the region, the team fed this data to advanced machine learning algorithms. Combining drone imagery, digital mapping and machine learning, the team configured region modeling and infrastructure development with higher accuracy. Another organization known as One Concern, a California-based startup, has created a predictive AI program called Seismic Concern to accurately predict seism and is also working on solutions for wildfires, floods and hurricanes.

Smart AI Agriculture

Another application of AI in developing countries is smart agriculture. Farmers monitor crops more effectively and make better predictions on planting, weeding and harvesting using AI tools. It can also be used to analyze one plant at a time and add pesticides only to infected plants and trees instead of spraying pesticides across large swaths of crops. One California-based tech company is an example of this use of AI. So, the developing countries farmers in rural parts of India are also using AI to increase yields through better access to information about the farming season than they would normally have. Technology-enabled process automation offers the agribusiness industry the chance for remarkable growth -- not only in developed countries but around the world. There's a unique opportunity to increase yields, cut down labor costs and improve people's health.

Medicine Delivery to developing countries' patients urgent need

Companies are also leveraging AI to improve access to health care in some of the most remote areas of the world. In Rwanda, for example, Zipline is using drones to deliver medical supplies and blood to hospitals and clinics that are difficult to access by car. This has dramatically impacted people living in remote parts of the country because they are able to get medical help when needed. The drone system in Rwanda has also helped reduce waste of blood by 95%, as noted by Zipline. One Concern has created an AI program called Seismic Concern that accurately predicts seismic events and is also working on solutions for floods, wildfires and hurricanes. The medical field may actually benefit the most from emerging technologies in developing countries.

Assistance to reduce teaching work workload or psychological pressure to teachers in developing countries' schools

Another vital area benefiting from innovative technologies like AI is education. Advanced technologies can enhance how we learn, teach and perform tasks. In most developing countries, schools lack experienced teachers and resources to enhance students' knowledge. As a result, many students still have to walk long distances to get to the nearest school, which has created education gaps, especially in rural areas. AI tools such as personalized learning assistants can simplify learning by making tutoring services and learning materials accessible to all students, wherever they are. Machines can be automated to help students learn basic concepts without a tutor, which companies like Carnegie Learning are working on. This would allow students to learn at any time from anywhere. With AI, education is made easy and accessible to more people.

The initial usage of AI in developing countries has been at a micro level -- solving small, specific problems in a defined industry. As machine learning advances and there is a higher utilization of AI, we will see more complex issues being targeted and resolved. When duly adopted, AI can positively impact future developing countries people everyday lives not just in disaster intervention, education, health care and agriculture but can also help in mitigating poverty, malnutrition and pollution. Especially, in developing nations, to leverage AI's true potential and create a snowball effect. Startups are defining a holistic and humanitarian approach to building more sophisticated, AI-ready societies. Stakeholders in the AI landscape should understand the strengths and nuances of the developing world as well as the limitations of AI and create localized solutions and applications.

Why does smart phone help developing countries communication ?

Internet Seen as Positive Influence on Education but Negative on Morality in Emerging and Developing Nations. Internet access differs substantially across the 32 emerging and developing countries polled, with the lowest rates of internet use in South Asian and sub-Saharan African nations. Within countries, computer owners, young people, the well-educated, the wealthy and those with English language ability are much more likely to access the internet than their counterparts. To access the internet, people increasingly use smartphones rather than more cumbersome fixed landline connections and computers. Around the world, both smartphones and basic-feature phones alike are used for sending messages and taking pictures.

In fact, many developing countries young people, students are popular to use smart phones for internet usage aim, instead of communication. Moreover, many developing countries working people are also popular to use smart phones for any working usage in their working time , even non working time any time. So, smart phones (AI) phones will be important communication or leisure tools to developing countries people in the future. Unless, it is one day, scientists can develop another new communication tool to replace smart phones. So, artificial intelligence will be important to influence developing countries people , how to improve or bring positive learning attitudes to students in their daily learnnng lifes. as well as how to raise developing countries people, how to raise working people efficiency or improve performace in their daily working lifes. So, AI may bring positive learning or working attitudes to developing countries working people and students both.

The Positive Impact of Mass Media in Developing Countries

Radio, newspapers, television, Internet, social media, etc., all of these are forms of mass media. Each of these outlets has the capability of bringing information to thousands of people with one device. While in some communities it is easy to take advantage of these communication outlets such as television and Internet access, not everyone has access to such outlets. Radio is one of the most common forms of mass media in developing countries because it's affordable and uses less electricity than many other forms of mass media, but only approximately 75 percent of people in developing countries have access to a radio, and roughly 77 percent of people in rural areas have access to electricity.

For developing countries that have implemented forms of mass media in their communities, there have been numerous positive outcomes are influenced to impact developing countries mass media by artificial intelligence as below:

When AI is participated to developing countries mass media, it can influence any radio, television audiences raise more attention to each other through social media platforms such as Facebook and Twitter and create, organize and initiate street protests and campaigns. Furthermore, having access to social media in developing countries, people are able to connect to those that they usually wouldn't have the chance to talk to. Moreover, AI Provides educational opportunities- In many countries, the division between local and national languages as well as issues of literacy can make communication difficult. With the use of mass media, a bridge can be built between these two gaps. In India, there is a radio station that provides information in local languages and respects local culture and traditions. One of the main ways is to create public awareness of what is going on with businesses and government officials. The media plays an important role in giving people the opportunity to act against injustice, oppression and misdeeds that they otherwise wouldn't know about. Information on available healthcare, a mass radio broadcast was sent out encouraging parents to seek treatment at local healthcare facilities for their sick children. With this mass outreach on healthcare, the encouragement of people to take their children to healthcare facilities saved thousands of lives. This easy way of encouraging others and bringing awareness about certain diseases was made possible through a simple radio broadcast. Finally, when AI is particiapted to media, it may bring many social issues to life that otherwise would remain unknown to many people. In developing countries and communities like Burkina Faso, when the radio broadcast was released about malaria, diarrhea and pneumonia, people were educated and moved to action and knew to take their children to healthcare facilities for preventative care. As it is seen, having access to different media outlets is vital for those in developing countries. Here are three ways that those in developing countries can implement mass media to help their people and communities.

When AI is participated to any internet radio or internet newspaper mass online listening or reading channel. It can provide online radios or newspapers in public places- By providing online radios and newspapers in public areas it gives community members to access news, information and emergency warnings. Even though radios can be on the cheaper side, there are still many people that can't afford to have a radio in their home. By providing one in a local place, not only would it better educate the community members but also it will bring the community together. So, it can make media outlets a two-way platform- Creating a two-way platform between the community and those who are behind the radio stations, newspapers or broadcasts makes the community feel involved and that their voices are being heard. An organization called Soul City in sub-Saharan Africa is showing how well two-way platforms work by engaging their listeners and having them contribute thoughts and ideas about complex issues. Because developing countries radio listening audiences or newspaper readers are popular to accept computer online radio listening channel or online newspaper reading channel to replace traditional paper newspapers or radio machines. So, AI may raise their listening news or reading news leisure feeling from online mass media channel in the future.

● Why do developed countries need to develop AI

Artificial intelligence, or AI, is driving massive shifts across the globe, and every day more questions arise. What impact will AI have on the workforce and how can we prepare for it? How can we encourage economy-boosting and job-creating technologies? How can we ensure that AI will be implemented ethically and with minimal bias? How will society benefit? For developed country, such as US example. None of the US, Israel and Russia have a formal national AI policy yet. Private sector companies such as Google, Amazon and Apple and the US department of defence are driving the bulk of AI investment in the United States. Though Israel does not have a specific policy, it is keenly focused on AI and has seen the number of AI start-ups triple since 2014.

Developed country may learn whether what weakness it is lacking when it does not continue to develop AI from one another developed country. Which countries are approaching AI most effectively, and to what degree is there opportunity for greater international collaboration? It may be too early to tell; however, when analyzing the best practices of existing national AI policies, there is much that can be learned. These are the specific areas to consider. When one developed country continue to develop or research AI, it may bring these benefits as below:

On gathering Data aspect, from self-driving vehicles to smart cities, data is the driver behind AI. Innovation in the United States is limited without a national strategy that answers questions about protocol and ownership. France and Denmark, on the other hand, are opening government data. France is hosting troves of centrally collected public and private data that it plans to make available as part of its strategy. Conversely, by taking a restrictive position on issues of data collection (as indicated by the implementation of General Data Protection Regulation), the EU is putting manufacturers and software designers at a disadvantage while balancing the demand for privacy. On raising technologica talent aspect, the demand for AI talent far outweighs the available supply. As a result, almost every nation's strategy addresses talent development. Canada's AI strategy is distinct in that it primarily focuses on research and talent strategy. The country boasts AI degree programmes and is building a $127 million research facility in Toronto. Companies like Facebook and my own company, Uptake, are investing in Canada to access this talent pool. On AI legal technological innovation aspect, a whole host of legal questions swirl around AI. The country is developing a bill for AI liability that will be ready in March 2019. The government hopes the legal framework will attract investors by providing a simple, comprehensive guideline to enable the broad use of AI systems. So, when the developed country applied AI technology to assist any lawyers to work, then AI can help them to reduce the workload to draft any legal documents more easier. So, any developed countries lawyers' draft legal documents time must reduce if the developed countries lawyers accept to apply AI to assist their legal works. One of the great promises of AI is its potential for improving quality of life. But without the right planning and oversight, we risk exacerbating problems of inequality or marginalizing groups of people. As an example, India's AI strategy is focused on leveraging the technology not only for economic growth, but also for social inclusion.

AI may bring what benefits to developed countries

From SIRI to self-driving cars, artificial intelligence (AI) is progressing rapidly. While science fiction often portrays AI as robots with human-like characteristics, AI can encompass anything from Google's search algorithms to IBM's Watson to autonomous weapons. Artificial intelligence today is properly known as narrow AI (or weak AI), in that it is designed to perform a narrow task (e.g.

only facial recognition or only internet searches or only driving a car). However, the long-term goal of many researchers is to create general AI (AGI or strong AI). While narrow AI may outperform humans at whatever its specific task is, like playing chess or solving equations, AGI would outperform humans at nearly every cognitive task.

Why research AI safety? Would AI bring war when AI is continued to develop by developed countries? In the near term, the goal of keeping AI's impact on society beneficial motivates research in many areas, from economics and law to technical topics such as verification, validity, security and control. Whereas it may be little more than a minor nuisance if your laptop crashes or gets hacked, it becomes all the more important that an AI system does what you want it to do if it controls your car, your airplane, your pacemaker, your automated trading system or your power grid. Another short-term challenge is preventing a devastating arms race in lethal autonomous weapons.

In the long term, an important question is what will happen if the quest for strong AI succeeds and an AI system becomes better than humans at all cognitive tasks. As pointed out by I.J. Good in 1965, designing smarter AI systems is itself a cognitive task. Such a system could potentially undergo recursive self-improvement, triggering an intelligence explosion leaving human intellect far behind. By inventing revolutionary new technologies, such a superintelligence might help us eradicate war, disease, and poverty, and so the creation of strong AI might be the biggest event in human history. Some experts have expressed concern, though, that it might also be the last, unless we learn to align the goals of the AI with ours before it becomes superintelligent.

There are some who question whether strong AI will ever be achieved, and others who insist that the creation of superintelligent AI is guaranteed to be beneficial. At FLI we recognize both of these possibilities, but also recognize the potential for an artificial intelligence system to intentionally or unintentionally cause great harm. We believe research today will help us better prepare for and prevent such potentially negative consequences in the future, thus enjoying the benefits of AI while avoiding pitfalls.

How can AI be dangerous when developed countries continue to develop AI to become weapon to replace soldiers?

Most researchers agree that a superintelligent AI is unlikely to exhibit human emotions like love or hate, and that there is no reason to expect AI to become intentionally benevolent or malevolent. Instead, when considering how AI might become a risk, experts think two scenarios most likely:

The AI is programmed to do something devastating: Autonomous weapons are artificial intelligence systems that are programmed to kill. In the hands of the wrong person, these weapons could easily cause mass casualties. Moreover, an AI arms race could inadvertently lead to an AI war that also results in mass casualties. To avoid being thwarted by the enemy, these weapons would be designed to be extremely difficult to simply "turn off," so humans could plausibly lose control of such a situation. This risk is one that's present even with narrow AI, but grows as levels of AI intelligence and autonomy increase.

The AI is programmed to do something beneficial, but it develops a destructive method for achieving its goal: This can happen whenever we fail to fully align the AI's goals with ours, which is strikingly difficult. If you ask an obedient intelligent car to take you to the airport as fast as possible, it might get you there chased by helicopters and covered in vomit, doing not what you wanted but literally what you asked for. If a superintelligent system is tasked with a ambitious geoengineering project, it might wreak havoc with our ecosystem as a side effect, and view human attempts to stop it as a threat to be met. So, a super-intelligent AI will be extremely good at accomplishing its goals, and if those goals aren't aligned with ours, we have a problem. You're probably not an evil ant-hater who steps on ants out of malice, but if you're in charge of a hydroelectric green energy project and there's an anthill in the region to be flooded, too bad for the ants. A key goal of AI safety research is to never place humanity in the position of those ants.

Why the recent interest in AI safety ?
Stephen Hawking, Elon Musk, Steve Wozniak, Bill Gates, and many other big names in science and technology have recently expressed concern in the media and via open letters about the risks posed by AI, joined by many leading AI researchers. The idea that the quest for strong AI would ultimately succeed was long thought of as science fiction, centuries or more away. However, thanks to recent breakthroughs, many AI milestones, which experts viewed as decades away merely five years ago, have now been reached, making many experts take seriously the possibility of superintelligence in our lifetime. While some experts still guess that human-level AI is centuries away, most AI researches at the 2015 Puerto Rico Conference guessed that it would happen before 2060. Since it may take decades to complete the required safety research, it is prudent to start it now.
Because AI has the potential to become more intelligent than any human, we have no surprise way of predicting how it will behave. We can't use past technological developments as much of a basis because we've never created anything that has the ability to, wittingly or unwittingly, outsmart us. The best example of what we could face may be our own evolution. People now control the planet, not because we're the strongest, fastest or biggest, but because we're the smartest. If we're no longer the smartest, are we assured to remain in control?
A captivating conversation is taking place about the future of artificial intelligence and what it will/should mean for humanity. There are fascinating controversies where the world's leading experts disagree, such as: AI's future impact on the job market; if/ when human-level AI will be developed; whether this will lead to an intelligence explosion; and whether this is something we should welcome or fear. But there are also many examples of of boring pseudo-controversies caused by people misunderstanding and talking past each other. When one developed country continue to develop AI, can itself country's all factories workers will lose jobs, due to AI can replace them to do simple works in factories, or any

public transport drivers, e.g. bus drivers, ferry , tram, train drivers, they will lose jobs, when AI (non manual driving drivers) can replace all public transport drivers. So, some occupations will lose if developed countries continue to develop or research AI to replace human to do some simple jobs, such as some cooking jobs can be done by AI. So, it is possible that future cookers won't be needed, because AI cooking skills may be better than them to cook any good taste chinese or western food in restaurants. If you drive down the road, you have a subjective experience of colors, sounds, etc. But does a self-driving car have a subjective experience? Does it feel like anything at all to be a self-driving car? Although this mystery of consciousness is interesting in its own right, it's irrelevant to AI risk. If you get struck by a driverless car, it makes no difference to you whether it subjectively feels conscious. In the same way, what will affect us humans is what superintelligent AI does, not how it subjectively feels.

In fact, AI may be make any brokers jobs in financial market. the main concern of the beneficial-AI movement isn't with robots but with intelligence itself: specifically, intelligence whose goals are misaligned with ours. To cause us trouble, such misaligned superhuman intelligence needs no robotic body, merely an internet connection – this may enable outsmarting financial markets, out-inventing human researchers, out-manipulating human leaders, and developing weapons we cannot even understand. Even if building robots were physically impossible, a super-intelligent and super-wealthy AI could easily pay or manipulate many humans to unwittingly do its bidding. So, future brokers will be replaced by AI, when AI can be made to own financial brokers' analytical mind to make more accurate whether the share price will rise up or fall down to compare human financial brokers' analytical mind. The robot misconception is related to the myth that machines can't control humans. Intelligence enables control: humans control tigers not because we are stronger, but because we are smarter. This means that if we cede our position as smartest on our planet, it's possible that we might also cede control.

Not wasting time on the above-mentioned misconceptions lets us focus on true and interesting controversies where even the experts disagree. What sort of future do you want? Should we develop lethal autonomous weapons? What would you like to happen with job automation? What career advice would you give today's kids? Do you prefer new jobs replacing the old ones, or a jobless society where everyone enjoys a life of leisure and machine-produced wealth? Further down the road, would you like us to create superintelligent life and spread it through our cosmos? Will we control intelligent machines or will they control us? Will intelligent machines replace us, coexist with us, or merge with us? What will it mean to be human in the age of artificial intelligence?

Why do developed countries people need AI ?

Why do we assume that AI will require more and more physical space and more power when human intelligence continuously manages to miniaturize and reduce power consumption of its devices. How low the power needs and how small will the machines be by the time quantum computing becomes reality? Why do we assume that AI will exist as independent machines? If so, and the AI is able to improve its Intelligence by reprogramming itself, will machines driven by slower processors feel threatened, not by mere stupid humans, but by machines with faster processors? What would drive machines to reproduce themselves when there is no biological incentive, pressure or need to do so?

Who says superior AI will need or want to have a physical existence when an immaterial AI could evolve and preserve itself better from external dangers. What will happen if AI developed by competing ideologies, liberalism vs communism, reach maturity at the same time, will they fight for hegemony by trying to destroy each other physically and/or virtually. If AI is programmed to believe in God, and competing AI emerges programmed by muslims, christians or jews, how are the different AI's going to make sense of the different religious beliefs, are we going to have AI religious wars? What if the "powers that be" greatest fear is the emergence of a super AI that police's and rationalizes the distribution of wealth and food.

A friendly super AI that is programmed to help humanity by, enforcing the declaration of Human Rights (the US is the only industrialized country that to this day has not signed this declaration) ending corruption and racism and protecting the environment.Most benefits of civilization stem from intelligence, so how can we enhance these benefits with artificial intelligence without being replaced on the job market and perhaps altogether?

Key to the process of machine learning are neural networks. These are brain-inspired networks of interconnected layers of algorithms, called neurons, that feed data into each other, and which can be trained to carry out specific tasks by modifying the importance attributed to input data as it passes between the layers. During training of these neural networks, the weights attached to different inputs will continue to be varied until the output from the neural network is very close to what is desired, at which point the network will have 'learned' how to carry out a particular task. A subset of machine learning is deep learning, where neural networks are expanded into sprawling networks with a huge number of layers that are trained using massive amounts of data. It is these deep neural networks that have fuelled the current leap forward in the ability of computers to carry out task like speech recognition and computer vision.

In conclusion, when developed countries continue to develop AI, it may bring positive advantages to bring raising productivies, or efficiencies, but it may also raise unemployment ratio to any low skill or low knowledge jobs in ther societies. However, human future society will need to change to be better to raise our living standard. But AI is one kind the best choice tool to achieve this aim in our future, so I agree developed countries continue to develop or research AI to be the super -human machine.

Reference

A. Castano et. al. " Automatic detection of dust devils and clouds at Mars" Machine vision and applications, Oct. 2008, vol. 19, no 5-6, pp. 467-482.

Accenture, " Why artificial intelligence is the future of growth"(2017) <http://www.accenture.com/us-en/insight-a rtificial-intelligence-future-growth>.

D. Schedidt , Unmanned Air Vehicle Command And Control, Handbook Of Unmanned Air Vehicles, Springer-Verlag, 2014. Facebook (AI) Research Available at https://research.facebook.com/ai, research at google, machine intelligence available at http://research.google.com/pubs/machineintellige nce.html; micro soft research-machine learning and artificial intelligence available at http://research.microsoft.com/en-us/research- areas/machine-learning-ai.aspx.

International Federation Of Robotics, 2016. IFR press release world robotics report. IFR, org . 29 Sept. Accessed Feb. 01, 2017. http://www.ifr.org/news/ifr-press-release/world-robitics report -2016-8321.

K, Fedra , "GIS and environmental modelling" in environmental modelling with GIS, edited by M.F. Goodchild.B.O. Parks and L.T. Steyaert, Oxford University press, pp. 35-50, 1994.

Keynes, J.M. (1933). Economic possibilities for our grandchildren (1930). Essays in persuasion, pp.358-73.

Mckinsey & Company (2013, May). Disruptive technologies: Advices that will transform life, business and the global economy , USA.

Ministry of economy, trade and industry, Japan, 2015, Japan's robot strategy. Ministry of economy, trade and industry.

Ray Kurzweil , The age of spiritual machines (1999) is cited numerously through this chapter: Kurzweilai.net http://www.kurzweilai.net

Rich, Elaine & Knight, Kevin, Artificial Intelligence Second Edition, 1991, New York; Mc-Graw-Hill.

Artificial intelligence bank service working environment

Focus on outcomes not technology.Artificial Intelligence: Waiting to be unleashed? The Insider Column - When Digital Transformation misses. Are you meeting the demands of the new digital consumer? Will your legacy mindset compromise your digital competitiveness? Can artificial intelligence create online remote office new business service market in global ?

The Future of Artificial Intelligence In The Workplace:

Is AI going to displace workers or come as a benefit to them?

Is AI going to displace workers or come as a benefit to them? Getty

Smart technologies aren't just changing our homes; they're edging their way into their numerous industries and are disrupting the workplace. Artificial Intelligence (AI) has the potential to improve productivity, efficiency and accuracy across an organization – but is this entirely beneficial? Many fear that the rise of AI will lead to machines and robots replacing human workers and view this progression in technology as threat rather than a tool to better ourselves.

With AI continuing to be a prominent online office service business to replace human actual office working environment, businesses need to realize that self-learning and black-box capabilities are not the panacea. Many organisations are already beginning to see the incredible capabilities of AI, using these advantages to enhance human intelligence and gain real value from their data. As there is increasing evidence demonstrating the benefits of intelligent systems, more decision-makers in the boardroom are gaining a better understanding of what AI can really offer. Research conducted by EY explains "organizations enabling AI at the enterprise level are increasing operational efficiency, making faster, more informed decisions and innovating new products and services." Can articial intelligent technology create remote office working environment to replace our traditional actual office work environment ? Can we do not need to go to office to work , when any office staffs ,e.g. managers, clerk, etc. they can apply artificial intelligent technology and online technology to work at home, such as remote office working environment ?

The first companies employing AI systems across the board will gain competitive advantage, reduce cost of operations and remove head counts. Whilst this may be a positive from a business perspective, it is obvious why this a worry for those working in roles at risk of displacement. The introduction of these technologies will likely trigger an issue with unions and job security due to the substantial operational changes. Although AI will affect every sector in some way, not every job is at equal risk. PwC predicts a relatively low displacement of jobs (around 3%) in the first wave of automation, but this could dramatically increase up to 30% by the mid-2030's. Occupations within the transport industry could potentially be at much greater risk, whereas jobs requiring social, emotional and literary abilities are at the lowest risk of displacement.

A positive future with artificial intelligence to bring remote online office working environment chance:

Many businesses and individuals are optimistic that this AI-driven shift in the workplace will result in more jobs being created than lost. As we develop innovative technologies, AI will have a positive impact on our economy by creating jobs that require the skill set to implement new systems. 80% of respondents in the EY survey said it was the lack of these skills that was the biggest challenge when employing AI programs.

It is likely that artificial intelligence will soon replace jobs involving repetitive or basic problem-solving tasks, and even go beyond current human capability. AI systems will be making decisions instead of humans in industrial settings, customer service roles and within financial institutions. Automated decisioning will be responsible for tasks such as approving loans, deciding whether a customer should be onboarded or identifying corruption and financial crime.

Organisations will benefit from an increase in productivity as a result of greater automation, meaning more revenue will generated. This thus provides additional money to spend on supporting jobs in the services sector.Due to the vast array of jobs that could be impacted by AI, it is fundamental to address the potential pitfalls of these technologies. Business need to overcome the trust and bias

issues surrounding AI by achieving an effective and successful implementation that makes it possible for everyone to benefit.
Governments must ensure that gains from AI are shared widely across society to prevent social inequality between those affected and unaffected by these developments. For example, this could be through increased investment into training. With the additional cost-savings from implementing AI systems, employers should also focus on upskilling their current employees.
To properly leverage the power of AI, we need to address the issue at an educational level, as well as in business. Education systems needs to focus on training students in roles directly associated to working with AI, including programmers and data analysts. This requires more emphasis to be put on STEM subjects (science, technology, engineering and mathematics). Also, subjects centered around building creative, social and emotional skills should be encouraged. Whilst artificial intelligence will be more productive than human workers for repetitive tasks, humans will always outperform machines in jobs requiring relationship-building and imagination. Hence, artificial intelligence will change our world both inside and outside the workplace. Instead of focusing on the fear surrounding automation, businesses need to embrace these new technologies to ensure they implement the most effective AI systems to enhance and compliment human intelligence.

Artificial Intelligence (AI) in Banking working environment

Artificial Intelligence (AI) is a fast-evolving technology, gaining popularity all around the world. Several industries have already adopted AI for various applications, getting better and smarter day by day. In the past few years, the banking sector has also become one of the leading adopters of Artificial Intelligence. Most banks and financial institutions are implementing AI to add more efficiency to their back-office and lessen security risks.
As per Statista, the AI market in the United States is forecasted to reach 7.35 billion U.S. dollars in 2018. Some major applications of AI include classification, image recognition, object identification,

and automated geophysical feature detection. Speaking of banking and financial institutions, JPMorgan Chase, Wells Fargo, Bank of America, CitiBank, and other leading U.S. banks have already implemented AI in their systems, helping consumers manage their daily banking needs more efficiently.

AI technology can bring better Customer Support in bank service environment

Several pieces of evidence advocate that the customers willingly prefer self-service options which allow them to chat with a virtual assistant as if it were a live customer representative. Most leading banks have already added virtual assistants to their instant website chatbots, voice response systems, and mobile applications. Artificial Intelligence considers each interaction as a teachable moment, so the chatbots (virtual assistants) keeps getting better while understanding customers. With AI, virtual assistants can deliver better customer support. It also allows sentiment analysis, so the virtual assistant can determine when individuals are getting frustrated and instantly transfer them to a live agent.

Enhanced Banking Services

AI streamlines the banking process while giving customer service a new level of comfortability. It allows banks to meet customers' expectations with comprehensive digital support. With Artificial Intelligence, you can achieve greater precision and accuracy. From cash transfer to bills payment, cards management, and other support, AI can significantly enrich the satisfaction level of your customers. All of these operations can be easily managed through desktops, smartphones, and other mobile devices.

Scam Recognition

With an immense growth of banking fraud, scam recognition and reduction has become challenging for the banking sector. Several banks tried to identify the factors and powerful solutions but couldn't succeed. However, AI makes it easier to detect the factors involved in frauds and support investigators. It improves financial security with advanced fraud prevention tactics. Artificial Intelligence works as a real-time scam solution for the banking sector while handling complex situations and tactics. Based on advanced data crunching, AI can detect fraud by flagging unusual transactions. It also feeds back into the consumer's profile which subsequently builds a secure environment.

Advanced Data Analytics

One of the main advantages of AI is its ability to complete tedious tasks through intricate automation, resulting in better productivity. Based on a machine learning algorithm, AI can quickly consume and process a massive amount of data at an expedited level. The enormous speed brings efficiency to financial services, providing scope for personalized offerings to consumers. What's even more, AI makes faster decisions while carrying out actions quickly. With such advantages, it is nearly obvious that the majority of banks and financial institutions will adopt AI to stay competitive and deliver better customer support. However, several cons are also associated with a machine learning algorithm. As it continues to learn and grow, the decision-making capabilities may create problems in the near future.

Disadvanages of AI in Banking Sector

Artificial intelligence is also expected to massively disrupt banks and traditional financial services. Some of its disadvantages are listed below.

Highly Expensive

Production and maintenance of artificial intelligence demand huge costs since they are very complex machines. AI also consists of advanced software programs which require regular updates to meet the needs of the changing environment. In the case of critical failures, the procedure to reinstate the system and recover lost codes may require enormous time and cost.

Bad Calls

Though Artificial Intelligence can learn and improve, it still can't make judgment calls. Humans can take individual circumstances and judgment calls into account when making decisions, something that AI might never be able to do. Replacing adaptive human behavior with AI may cause irrational behavior within ecosystems of humans and things.

Distribution of Power

There is a constant fear of AI superseding or taking over the humans. Artificial intelligence can give a lot of power to the few individuals who are controlling it. Hence, AI carries the risk and takes control away from humans while dehumanizing actions in several ways.

Unemployment

Replacement of the workforce with machines can lead to wide-reaching unemployment. Moreover, if the use of AI becomes rampant, people will be highly dependent on the machines and lose their creative power. Unemployment is a socially undesirable issue. Individuals with nothing to do can lead to the devastating use of their minds. Be it banking or any other sector; Artificial intelligence can effectively increase the unemployment rate.

Artificial Intelligence delivered to wrong hands can turn out to be a serious threat to humankind. If individuals start thinking destructively, they can generate havoc with these advanced machines. The challenges introduced by the emergence of artificial intelligence revolve around several things. However, AI is a right balance of skill and emotions which is continually growing. Artificial intelligence provides banks, financial institutions, and tech companies with significant competitive advantages. Nevertheless, it can completely transform the financial sector and make it faster, but this will only be possible if the financial industry can manage the security risk of systems based on AI.

What does artificial intelligence mean for the bank service office workers?

With all these new artificial intelligence use cases comes the question of whether machines will force humans into obsolescence. The jury is still out: Some experts vehemently deny that artificial intelligence will automate so many jobs that millions of people find themselves unemployed, while other experts see it as a pressing problem.

"The structure of the workforce is changing, but I don't think artificial intelligence is essentially replacing jobs in bank service working environment. It allows us to really create a knowledge-based economy and leverage that to create better automation for a better form of life. It might be a little bit theoretical, but I think if you have to worry about artificial intelligence and robots replacing some bank service jobs, e.g. bank security, bank enquiry service,. But, AI can not replace bank counter service staffs to do saving or withdrawing money transfer tasks when any customers prepare to save money or withdraw money in bank counters. As this technology develops, the AI bank service will see new startups, numerous saving or withdraw transactions from consumer won't be raise more easily.

AI to Banking and Finance industry

The banking and finance industry plays a major role in our lives. I mean the world runs on money and banks are essentially the gatekeepers that regulate that flow. Did you know that the banking and finance industry heavily relies on artificial intelligence for things like customer service, fraud protection, investment, and more? A simple example is the automated emails that you receive from banks whenever you do an out of the ordinary transaction. Well, that's AI watching over your account and trying to warn you of any fraud.

AI is also being trained to look at large samples of fraud data and find a pattern so that you can be warned before it happens to you. Also, when you hitch a little snag and chat with bank's customer service, chances are that you are chatting with an AI bot. Even the big players in the finance industry use AI to analyze data to find the best avenues to invest money so they can get the most returns with the least risk. That's not all, AI is poised to play an even bigger role in the industry as major banks across the world are investing billions of dollars in the AI technology and we all will observe its effects sooner than later

How AI influences our daily working life in any office working environment

Can AI bring only disadvantages? If AI can bring disadvantges, what are its disadvantages to any working environment ?The entire tech world is debating the consequences of artificial intelligence and the part AI is going to play in shaping our future. While we might think that artificial intelligence is at least a few years away from causing any considerable effects on our lives, the fact remains that it is already having an enormous impact on us. Artificial intelligence is affecting our decisions and our lifestyles every day. Don't believe me? I shall indicate some product examples how AI anticipates which can influence our working culture in any office environment. Examples of how Artificial Intelligence assistance to office working environment may include as below:

1. Smartphones

Smartphones have become the most indispensable tech product that we own today and we use it almost all the time. Well, if you are using a smartphone, you are interacting with AI whether you know it or not. From the obvious AI features such as the built-in smart assistants to not so obvious ones such as the portrait mode in the camera, AI is impacting our lives in every day office working environment.

In fact, the two examples that I provided that our working world of AI and how it is effecting our working lives. Firstly, there are the obvious AI elements which most of us have some knowledge about. For example, when you are using a smart assistant in office, whether it's Google Assistant, Alexa, Siri, or Bixby, you more or less know that these assistants are based on AI. However, when we are using a feature such as the portrait mode effect while shooting a picture, we never consider that AI might be behind that too. Have you ever thought how the Google Pixel phones or iPhones can capture such great portrait shots? The answer is artificial intelligence. So, when any office workers need to find any knowledge to solve their working problem immediately in any offices. They may apply AI smart phone tools to help them to apply online channel to search any new knowledge to attempt to solve their working problem in possible, when their computers have none any computers in offices.

Now more and more manufacturers are including AI in their smartphones with big chip manufacturers including Qualcomm and Huawei producing chips with built-in AI capabilities. The AI integration is helping in bringing features like scene detection, mixed and virtual reality elements, and more. AI is going to play an even major role in the coming years. We are already seeing the huge emphasis on AI with the latest Android and iOS updates. Features like app actions, splices, and adaptive battery in Android Pie and Siri shortcut and Siri suggestions in iOS 12 are made possible with AI. So, next time if any office workers think AI is not effecting them, take out your smartphone to replace computers to find any knowledge to help you to solve any tasks problems immediately in offices.

2. Social Media Feeds

If you are thinking that smart cars don't personally effect you as they are still not in your country or city, well, how about something which you use on a daily basis. Even if you are living under a rock, there's a high probability that you are tweeting from underneath it. If Twitter's not your choice of poison, maybe it's Facebook or Instagram, or Snapchat or any of the myriad of social media apps out there. Well, if you are using social media, most of your decisions are being impacted by artificial intelligence. So, any office workers may apply AI to help them to gather any new information to solve any difficult task problems , if their managers can not assist them to solve any sudden tasks problem, they are encountering to need to solve any working complex tasks problem internet social media in any any office working environment immediately.

From the feeds that office staffs can see in their working timeline to the notifications that you receive from these apps, everything is curated by AI. AI takes all your past behavior, web searches, interactions, and everything else that you do when you are on these websites and tailors the experience just for you. The sole purpose of AI here is to make the apps so addictive that you come back to them again and again, and I am ready to place a bet that AI is winning this war against you.

3. Online Ads Network

One of the biggest users of artificial intelligence is the online ad industry which uses AI to not only track user statistics but also serve us ads based on those statistics. Without AI, the online ad industry will just fail as it would show random ads to users with no connection to their preferences what so ever. AI has become so successful in determining our interests and serving us ads that the global digital ad industry has crossed 250 billion US dollars with the industry projected to cross the 300 billion mark in 2019. So next time when any product developers are going online and seeing ads or product recommendation, know that AI is impacting to any new products advertisement method more efficiently.

4. AI can be any office security

While we can all debate the ethics of using a broad surveillance system, there's no denying the fact that it is being used and AI is playing a big part in that. It is not possible for humans to keep monitoring multiple monitors with feeds from hundreds if not thousands of cameras at the same time, and hence, using AI makes perfect sense. With technologies like object recognition and facial recognition getting better and better every day, it won't be long when all the security camera feeds are being monitored by an AI and not a human. While there's still time before AI can be fully implemented such as security in any offices, this is going to be our future.

5. Smart Keyboard Apps

Smart Keyboard Apps. Granted, not everyone loves dealing with on-screen keyboards. However, they have become far more intuitive, allowing users to type comfortably and faster. What has probably proved to be a catalyst for them is the integration of AI. The smart keyboard apps keep a tab on the writing style of a user and predict words and emojis accordingly. Thus, typing on the touchscreen has become faster and more convenient. Not to mention, artificial intelligence also plays a vital role in pin-pointing misspellings and typos. So, any office workers can apply smart keyboard apps to help their to raise typing efficiency and reduce wrong typing word in error when they need to type any document in offices.

6. E-Commerce

` AI-driven algorithms have kind of given the much-needed impetus to e-commerce to provide a more personalized experience. According to several reports, its usage has vastly increased sales and also played a good part in building loyal relationships with customers. Thus, companies take advantage of AI to deploy chatbots to collect pivotal data and also predict purchases to create a customer-centric experience. Yet to come across this shift of strategy? Just spend some time with sites like Amazon and eBay and you will soon get to know how fast the landscape is changing around you – for the better! So, Ai can help any businesses to achieve e-commerce sale channel more easily.

7. Smart Email Apps

In any office working environment, if you still find your inbox cluttered with too many unwanted messages, chances are pretty high that you are still stuck with an old school email app. You heard it right! Modern email apps like Spark make the most of AI to get rid of spam messages and also categorize emails so that you can quickly access the important ones. What's more, they also offer smart replies based on the messages you receive to help you reply to any email quickly. The "Smart Reply" feature of Gmail is a great example of this. It uses AI to scan the text of the email and provides you with contextual answers. So, AI can help any office staffs to know who had sent any message from email and respond their email immediate , when AI can help any offices to avoid to receive any email spam rubbish email message in any time, even after working hours, it means that AI is working to help any office staffs to avoid to receive any email spam rubblish message in any time. So, when they go to office to work, even they go home after working hours. They can know whether what the important email messages are sent to their office email in boxes any time. Then, they can send email to respond their customers' enquires any time. So, AI can help any office workers can have chance to work at homes.

The Future of Artificial Intelligence In The Workplace
Smart technologies aren't just changing our homes; they're edging their way into their numerous industries and are disrupting the workplace. Artificial Intelligence (AI) has the potential to improve productivity, efficiency and accuracy across an organization – but is this entirely beneficial? Many fear that the rise of AI will lead to machines and robots replacing human workers and view this progression in technology as threat rather than a tool to better ourselves.

With AI continuing to be a prominent buzzword in 2019, businesses need to realize that self-learning and black-box capabilities are not the panacea. Many organisations are already beginning to see the incredible capabilities of AI, using these advantages to enhance human intelligence and gain real value from their data. As there is increasing evidence demonstrating the benefits of intelligent systems, more decision-makers in the boardroom are gaining a better understanding of what AI can really offer. Research conducted by EY explains "organizations enabling AI at the enterprise level are increasing operational efficiency, making faster, more informed decisions and innovating new products and services."

Today In: Cybersecurity

The first companies employing AI systems across the board will gain competitive advantage, reduce cost of operations and remove head counts. Whilst this may be a positive from a business perspective, it is obvious why this a worry for those working in roles at risk of displacement. The introduction of these technologies will likely trigger an issue with unions and job security due to the substantial operational changes. Although AI will affect every sector in some way, not every job is at equal risk. PwC predicts a relatively low displacement of jobs (around 3%) in the first wave of automation, but this could dramatically increase up to 30% by the mid-2030's. Occupations within the transport industry could potentially be at much greater risk, whereas jobs requiring social, emotional and literary abilities are at the lowest risk of displacement.

A positive future with artificial intelligence

Many businesses and individuals are optimistic that this AI-driven shift in the workplace will result in more jobs being created than lost. As we develop innovative technologies, AI will have a positive impact on our economy by creating jobs that require the skill set to implement new systems. 80% of respondents in the EY survey said it was the lack of these skills that was the biggest challenge when employing AI programs. It is likely that artificial intelligence will soon replace jobs involving repetitive or basic problem-solving tasks, and even go beyond current human capability. AI systems will be making decisions instead of humans in industrial settings, customer service roles and within financial institutions. Automated decisioning will be responsible for tasks such as approving loans, deciding whether a customer should be onboarded or identifying corruption and financial crime. Organisations will benefit from an increase in productivity as a result of greater automation, meaning more revenue will generated. This thus provides additional money to spend on supporting jobs in the services sector.

How to take advantage of AI to any offices

Due to the vast array of jobs that could be impacted by AI, it is fundamental to address the potential pitfalls of these technologies. Business need to overcome the trust and bias issues surrounding AI by achieving an effective and successful implementation that makes it possible for everyone to benefit. Governments must ensure that gains from AI are shared widely across society to prevent social inequality between those affected and unaffected by these developments. For example, this could be through increased investment into training.With the additional cost-savings from implementing AI systems, employers should also focus on upskilling their current employees.

To properly leverage the power of AI, we need to address the issue at an educational level, as well as in business. Education systems needs to focus on training students in roles directly associated to working with AI, including programmers and data analysts. This requires more emphasis to be put on STEM subjects (science, technology, engineering and mathematics). Also, subjects centered around building creative, social and emotional skills should be encouraged. Whilst artificial intelligence will be more productive than human workers for repetitive tasks, humans will always outperform machines in jobs requiring relationship-building and imagination. Artificial intelligence will change our world both inside and outside the workplace. Instead of focusing on the fear surrounding automation, businesses need to embrace these new technologies to ensure they implement the most effective AI systems to enhance and compliment human intelligence

How AI can help office workers to do tasks more easily

Companies are currently spending big on artificial intelligence and machine learning initiatives to the tune of $12 billion, but estimates put that figure as high as $57.6 billion by 2021, according to the International Data Corporation (IDC). With such massive shifts, the focus is usually on what we might lose, but it shouldn't be. A recent report on the future of work from the McKinsey Global Institute suggests that while only about 5% of jobs can be completely eliminated by automation, the rise of AI requires workers to beef up both technical and soft skills in order to stay competitive.

What's seldom discussed is how AI can revolutionize our jobs. It's now possible to pinpoint peak productivity for a single day, improve communication in meetings (even before people ever work together face to face), or even teach you to be a better leader, all thanks to AI platforms. I shall indicate these advantages to bring any office benefits from AI assistance as below:

1. AI can help any companies to get better to hire the best applicants

AI has the greatest potential to change the way companies find candidates, according to Alexander Rinke, cofounder and CEO of Celonis. The company's process-mining technology helps businesses to understand the areas where automation can help humans, he says. In HR departments, Celonis can help identify how fast workers come and go, the cost per hire, and which positions take the longest to fill. AI helped enable one customer's ability to identify bottlenecks in recruitment and reduced process costs internally by 30% as well as get them hired more quickly, he says.

Crafting a resume has never been easier, nor has landing an interview. Another example is how recruitment software provider iCIMS, in partnership with Google, is helping job seekers find jobs directly through the search engine, thanks to Google's AI and machine learning capabilities. Susan Vitale, iCIMS's chief marketing officer says that in addition to reducing the number of expired job postings, machine learning is underlying a private beta program of Google's Cloud Jobs Discovery model. "For a candidate searching for, say, a CTO role, Cloud Job Discovery will serve up CTO positions as well as jobs with titles that are similar, but not verbatim, such as chief technology officer or chief technical officer," says Vitale. This model also allows for conceptual search results, such as serving up job listings for cashiers, sales associates, and store associates when someone searches for one versus just only showing jobs that exactly match the keyword search criteria, she adds.

2. AI can help any office workers to raise much more productive efficiencies

John Furneaux, CEO and cofounder of Hive, says predictive analytics will help us better understand how we work. "It can tell us just about everything we want to know about teams and collaboration, for example, if men or women get more done in the afternoon, and if summer Fridays are a myth," he says. (Everyone thinks summer Fridays aren't productive, but in reality there's no difference between those and other Fridays during the year–productivity is equally low.)

Using a data set of over 30,000 completed actions across Hive workspaces, Furneaux says they were able to identify some notable trends in productivity. For example, men were far more productive early in the day, with a sharp decline in the afternoon, while women had a slower start to the day but were far more productive in later hours than their male counterparts. And analyzing chat messages revealed that women appear to complete more tasks when chatting, suggesting they use communication as a key tool to completing work. Similarly, Nintex Hawkeye analyzes data on business processes by types, users, roles, and departments to see who's doing the work and how long it takes them to do it. Management can monitor and analyze those metrics in real time.

3. AI can help any managers to make the most fair compensation and eliminate wage gaps to every staffs

Tanya Jansen, cofounder of the compensation management platform beqom, says that AI and predictive analytics can eliminate unconscious bias from compensation. Jansen says that AI based on a variety of rules including education, experience, certifications, and more can make compensation more fair and help businesses move closer to closing pay gaps. "Specifically, AI can help solve gender pay gaps and the CEO-to-worker pay gap, in which pay ratios of Fortune 500 companies range from 2:1 at the low end to nearly 5000:1 at the high end," she says. Additionally, the use of AI-driven compensation technology to make pay more fair can mitigate the risk of employee turnover, which costs businesses as much as 33% of a worker's annual salary to replace them.

4. AI can help any office staffs to arrange better meetings

Augmented Reality (AR) is still in its infancy, but AI and machine learning are the core components that make it work. As such, Christa Manning, the vice president and solution provider research leader at Bersin, Deloitte Consulting LLP, says that AR can help workers find the right information, in the right place, at the right time to make the best decisions wherever they may be working. For example, as more companies adopt video meetings and collaborative workspaces, it's likely we'll begin to see HR-curated information like talent profiles and work styles layered over interactions through AR."Imagine being in a video conference with a colleague and having direct insight into their communication style, seeing tips on how to best interact with them or reminders of what needs to be discussed. SO, AI can help any organizations to conclude or find the best methods to solve any problems after their every discussion in any meetings.

How AI is improving onboarding and training. AI coaching tools first learn by observing how different employees conduct specific tasks. Then these tools can walk new employees through how to complete those tasks—or even coach existing employees on how to do things more effectively or efficiently. Chorus is a great example of this technology. It analyzes sales calls while they happen, offering tips to help sales reps manage the cadence of meetings and use the most effective messaging. It also records all sales calls and compiles statistics for each sales rep, providing everyone with the tools they need to help them close more deals and conduct more effective calls. Another example is Cogito, a tool that combines AI with behavioral science to help customer service employees provide better phone support. It monitors calls for voice signals, providing real-time suggestions to representatives on how to improve the conversation.

5. AI can help any managers to be better leaders

Indiggo, a platform powered by a proprietary AI tool called "indi," functions as a brain that has consumed all the knowledge the company has gathered in its 15 years of operation. It also uses an algorithm to provide an estimate of how much time is wasted by a company by analyzing the size of its management team. Then it taps their calendars to see how they spend their time, and walks individual managers through a type of Q&A to make sure they are clear on what their top three priorities are, and how that relates to the organization's priorities, which will indicate if that strategy is moving forward or not. "The counterintuitive impact of these advances is that they actually make human work truly irreplaceable," Alexander Rinke, the cofounder and CEO of Celonis says. As such, he reminds us, "Humans are much better at processes that involve reasoning, judgment, and interaction with people." So, AI can recommend more accurate and useful opinions to help any managers to solve their managing challenges in office any time.

How AI is eliminating repetitive administrative tasks

There are a lot of tasks that knowledge workers spend time on that provide little—if any—value.For example, say you need to schedule a meeting to get consensus on a decision before moving forward, but you need five people to join the meeting. It's easy to spend a ton of time sending email back-and-forth or finding an open slot on everyone's calendar.That's not the most rewarding use of your time for you or your company.Tools like X.ai give employees AI-powered personal assistants that perform administrative tasks like scheduling, rescheduling, and cancelling meetings.

How AI is transforming internal communications and support

Personnel on the teams that provide employee support have their hands full with other responsibilities, too. HR teams work on building the kind of company people love working for. IT maintains the company's network and keeps data secure. Office managers frequently run big events like holiday parties.These tasks are crucial, but they're often hard for teams to focus on because they're busy answering routine questions. AI service desks like askSpoke allow employee support teams to balance their service commitments with other important responsibilities by reducing interruptions from rote, repetitive requests.Employees can askSpoke for whatever they need over Slack, email, SMS, and the web. askSpoke's friendly AI will automatically provide a prompt response.

How AI is transforming marketing, sales, and customer service

AI-powered chatbots help with external support as well. Just like with internal support tools like askSpoke, these chatbots learn from real marketers, salespeople, and customer service reps and are eventually able to answer questions as accurately as a knowledgeable person.For example, chatbot for Messenger helps customers plan their vacations. It books flights, hotels, and cars, highlights destination attractions, and even provides answers to questions like "Where can I go for $100 expense budget only?"

How AI is transforming business data and analytics

It's hard to run a competitive business today without data. But even massive amounts of data are useless without a way to transform that data into valuable insights. That's typically why you'd want to hire a data scientist—which just happens to be one of the most difficult roles to fill. How AI is fighting fraud and transforming security. Have you ever taken a call from your bank to find that someone used your debit card fraudulently? Most likely, your bank used some form of AI to detect the fraudulent transaction and decline it. Applying the same basic technology to the workplace helps identify security risks and keeps customer, employee, and company data safe. AI-powered software can automatically detect and address threats among thousands or millions of signals that humans would never be able to parse (especially not in real-time).

How AI is transforming productivity

While AI is transforming the workplace in many different ways across every industry, it's impacting productivity most of all. When your office staffs don't have to scroll through calendars to look for open meeting times, build reports in spreadsheets to look for insights, or spend your day answering the same questions over and over again, you're more productive. Workers are freed from redundant and mindless tasks, giving them more time to do work that matters, solve problems, and exercise their creativity. Some tools use AI to specifically monitor and boost productivity. For example, Deloitte's LaborWise provides company leaders and managers with productivity analytics that help them identify areas where labor costs are too high, impediments that slow people down, and departments that need additional staff.

In conclusion, what AI means for the workplace of the future. While some will dramatize the negative impacts of AI, cognitive computing, and robotics, these powerful tools will also help create new jobs, boost productivity, and allow workers to focus on the human aspects of work. Essentially, automation frees companies and their employees up to be more empathetic, to focus on things like the customer experience, employee engagement, and workplace culture.

What are traditional office tools to be replaced by AI ?
Artificial intelligence (AI) is predicted to eliminate over a million jobs in the next few years, potentially replacing lower level positions like administrative assistants with humanoid robots or voice assistants. But in the nearer future, fresh AI-driven software and products are also moving to eliminate non-human elements of the workplace by replacing traditional office tools, including both physical products and everyday electronic processes. Why should businesses switch from the tried-and-true to emerging technology? Many of the experts TechRepublic talked to said the AI options streamline business practices, making their adopters work smarter instead of harder. I shall indicate these office tools ,they can be applied to help any office staffs to finish their these tasks in office, they may include as below:

1. Scheduling

Workloud's end-to-end, cloud-based workforce management software takes scheduling from paper or Excel and moves it to the cloud. Everything from clocking in and out to monitoring employee absences is fully digitalized.Schedules and timesheets are accurate, created easily, and accessible through the service's web, tablet, and mobile apps. The software can also be used for absence management.

2. Employee talent selection

Using AI and organizational behavior science, can be used to replace internal spreadsheets and databases designed to monitor human capital. By mining employee attributes and experiences, the software can recommend who would be best for a project. The software also collects reviews after projects to better predict successful employee-project matches.The traditional hiring process is slow, biased and inaccurate, By removing humans from the beginning stages of the process, it can become faster and more fair, and result in better hires.
AI software automates the hiring process, using online simulations instead of manual screenings and interviews. Using the software, employers can include tasks in a job application, allowing job

candidates to show technical skills that may be necessary for a job. Employers can't rule out candidates until they see how the candidate performs, eliminating bias that occurs in the resume reading stage. Both sides also automatically receive updates about each other's steps, reducing the amount of time it takes to .

3.Timesheets: Allocate

Using AI and machine learning, the software registers an employee's computer activity throughout the day. The data, which can also pull information from email and calendars, is used to suggest timesheet entries to reflect a more accurate amount of time an employee spent working. The employee can review and revise as necessary. However, the software doesn't spy on or monitor employees. The data is only available to each employee, while others in the company can only see the timesheet's output, which Allocate said would be the same information available if a manual sheet was used. So,replacing manual timesheets with Allocate has three advantages: More accurate time entry, project analytics, and "'unsucking' the work experience."

4. Document storage

By using AI to read and analyze business and legal documents, AI can store all of the important document-based information in the cloud. The severe reduction in print-outs means less paper and ink, fewer products like binder clips and boxes to store and organize all of the paper, and more employee time freed up from not needing to manually sort through every document.

For example, in any lawyer offices, legal professionals' morale in the industry can suffer when they are pushed into performing such dull, repetitive tasks like sorting through and coding documents by hand, With AI tools to automate those duties, lawyers can focus on more meaningful projects and boost the business's and clients' success as a result. While focused on law firms, businesses that have a lot of unstructured data in documents may also be able to use the service to free up employee time and save on printing costs.

5. Scanners: Adobe Scan

While documents are moving to the cloud more and more,

sometimes a physical copy of a document still needs to be scanned using a bulky office scanner. Adobe Scan, an app that condenses a scanner to the size of a smartphone, can rid offices of the need for an in-house scanner. Users can download and open the app, then hold their device over whatever they need to scan. Adobe Sensei then turns the scan into a PDF, and sends it to the Adobe Document Cloud. The app can transform any image into digital text that can then be searched and used electronically. The app streamlines the scanning process, making scans cleaner and more immediate. For businesses already using Adobe services, the app makes documents easily accessible.

6. Landline phones

While landlines in homes are increasingly less common, the same cannot be said for offices. But using chatbots and AI integrations, RingCentral is trying to replace traditional office landline phone systems. The platform offers over 100 integrations, including that AI landline phones can let employees check their voicemail, and a Gong.io option that listens to call recordings to find traits of successful employees than can be used in training. An add-on for Gmail lets users switch from emailing back and forth to a voice session without needing to look up contact information. AI landline phone is easy to adopt and use in the workplace, and is more customizable than standard phone systems, said David Lee, vice president of platform products. Compared to the traditional option, the cloud-based option is "future-proof.

How artificial intelligence can raise office efficiency

Artificial Intelligence is already impacting every industry through automation and machine learning, bringing concerns that AI is on the fast track to replacing many jobs. But these fears aren't new, says Dan Jackson, director of Enterprise Technology at Crestron, a company that designs workplace technology. "I'd argue this is no different than when we moved from an agricultural to an industrial economy at the turn of the last century. The percentage of people working in agriculture significantly decreased, and it was a big shift, but we still have plenty of jobs 100 years later," he says.

Anytime society experiences a major technological advancement, we need to be prepared for it to change the way we live and work. It's hard to imagine what the future of jobs will look like with AI, but that future exists. And optimists suggest that, like the sewing machine to the textile industry, AI will make us better, more efficient and faster workers.

In fact, many experts agree that AI has the potential to eliminate mundane, administrative work, while we will always rely on human workers to be empathetic, collaborative, creative and strategic. But it's impact on any industry lies in the hands of the business leaders who are responsible for adopting AI strategies.

● Training presents challenges

A recent study of 1,000 global companies by Accenture found that AI is already creating three new categories of jobs: trainers, explainers and sustainers. Trainers are the people who teach AI systems how to act -- whether it's language, human behavior or the intricacies of human interaction. Explainers are the liaison between technology and business leaders, providing more insight and clarity into machine learning for the non-tech workers. Sustainers are the workers required to maintain AI systems and troubleshoot any potential issues. Some jobs were highly technical and required advanced degrees, but other roles demanded innately human things such as empathy and interaction. Downstream jobs, such as those in sales, marketing, or service will change to take advantage of the insights from AI, but many of the core skills will remain. However, it might sound like any job related to AI will require years of technical knowledge, but that isn't the case. We've already seen a shift in tech hiring -- companies often need highly specific skill sets that are hard to find in potential candidates. As a result, more businesses are hiring employees with the right soft skills, and then training them in technical skills.

An office effort measured approach to AI

The real takeaway is that any approach to AI will need to consider the human aspect of every business. AI has great potential to increase efficiency and accuracy and it's already been proven in certain industries. For example, the use of AI In banking to identify and money laundering schemes. It's also improved healthcare by "increasing the speed and accuracy" of cancer diagnosistics. AI can also help reduce the cost and length of human trafficking investigations, a situation where time is precious. In these examples, AI hasn't replaced jobs, but has positively impacted efficiency.

Thus, we need to ensure our education system responds to equip young people with the appropriate skills and adaptability, while businesses and public organizations must invest in training. Perhaps most of all, we need to encourage imagination and willingness to experiment. The organizations that can innovate with AI will reap the benefits. Their growth will make them the primary source of future jobs. Companies have a choice when implementing AI. They can choose to effectively implement systems that make employee's lives easier and find creative ways to leverage the technology. It's up to employers to ease fears for workers around AI and build strategies that benefit everyone. Hence, some AI experts believe AI can only raise efficiency to some office tasks, however, AI can not still raise efficiency to all office tasks for any office deparments. The reasons are because some office tasks which can only dominate to finish by human office workers. These office tasks are as below:

How can leaders and managers improve employee productivity while still saving time? These below tasks, AI experts ensure that AI can not help any office workers to raise their efficiencies as below:

1. Office managers can not delegate to AI to help them to do. While this tip might seem the most obvious, it is often the most difficult to put into practice. We get it–your company is your baby, so you want to have a direct hand in everything that goes on with it. While there is nothing wrong with prioritizing quality (it is what makes a business successful, after all), checking over every small detail yourself rather than delegating can waste everyone's valuable time. Instead, give responsibilities to qualified employees, and trust that they will perform the tasks well. This gives your employees

the opportunity to gain skills and leadership experience that will ultimately benefit your company. You hired them for a reason, now give them a chance to prove you right.

2. Office managers can not match Tasks to Skills to AI. Knowing your employees' skills and behavioral styles is essential for maximizing efficiency. For example, an extroverted, creative, out-of-the-box thinker is probably a great person to pitch ideas to clients. However, they might struggle if they are given a more rule-intensive, detail-oriented task. Asking your employees to be great at everything just isn't efficient–instead, before giving an employee an assignment, ask yourself: is this the person best suited to perform this task? If not, find someone else whose skills and styles match your needs.

3. Office managers can not teach AI to replace them how to communicate and teach their low level staffs how to work effectively. Every manager knows that communication is the key to a productive workforce. Technology has allowed us to contact each other with the mere click of a button (or should we say, tap of a touch screen)–this naturally means that current communication methods are as efficient as possible, right? Not necessarily. A McKinsey study found that emails can take up nearly 28% of an employee's time. In fact, email was revealed to be the second most time-consuming activity for workers (after their job-specific tasks). Instead of relying solely on email, try social networking tools (such as Slack) designed for even quicker team communication. You can also encourage your employees to occasionally adopt a more antiquated form of contact...voice-to-voice communication. Having a quick meeting or phone call can settle a matter that might have taken hours of back-and-forth emails. All of above communication tasks, I believe that AI can not do better than managers in offices.

4. AI can not keep Goals Clear and focused to be better than managers. You can't expect employees to be efficient if they don't have a focused goal to aim for. If a goal is not clearly defined and actually achievable, employees will be less productive. So, try to

make sure employees' assignments are as clear and narrow as possible. Let them know exactly what you expect of them, and tell them specifically what impact this assignment will have. One way to do this is to make sure your goals are "SMART" – specific, measurable, attainable, realistic, and timely. Before assigning an employee a task, ask yourself if it fits each of these requirements. If not, ask yourself how the task can be tweaked to help your workers stay focused and efficient.

5. AI can not know how to incentivize Employees to work more efficiently. One of the best ways to encourage employees to be more efficient is to actually give them a reason to do so. Recognizing your workers for a job well done will make them feel appreciated and encourage them to continue increasing their productivity. When deciding how to reward efficient employees, make sure you take into account their individual needs or preferences. For example, one employee might appreciate public recognition, while another would prefer a private "thank you." In addition to simple words of gratitude, here are a few incentives managers can know how to incentivize their staffs to work efficiently, but AI is only one machine, it can not perform very good.

6. AI does not know how to assist managers to train and Develop employees. Reducing training, or cutting it all together, might seem like a good way to save company time and money (learning on the job is said to be an effective way to train, after all). However, this could ultimately backfire. Forcing employees to learn their jobs on the fly can be extremely inefficient.

So, instead of having workers haphazardly trying to accomplish a task with zero guidance, take the extra day to teach them the necessary skills to do their job. This way, they can set about accomplishing their tasks on their own, and your time won't be wasted down the road answering simple questions or correcting errors. Past their original training, encourage continued employee development. Helping them expand their skillsets will build a much more advanced workforce, which will benefit your company in the long run. There are a number of ways you can support employee development: individual coaching, workshops, courses, seminars, shadowing or mentoring, or even just increasing their responsibilities. Offering these opportunities will give employees additional skills that allow them to improve their efficiency and productivity. But, AI do not know how to improve any office workers' performance more easily than managers.

- How can AI be dangerous to office working environment?

Most researchers agree that a superintelligent AI is unlikely to exhibit human emotions like love or hate, and that there is no reason to expect AI to become intentionally benevolent or malevolent. Instead, when considering how AI might become a risk to any office working environments, experts think two scenarios most likely:

The AI is programmed to do something devastating: Autonomous weapons are artificial intelligence systems that are programmed to kill. In the hands of the wrong person, these weapons could easily cause mass casualties. Moreover, an AI arms race could inadvertently lead to an AI war that also results in mass casualties. To avoid being thwarted by the enemy, these weapons would be designed to be extremely difficult to simply "turn off," so humans could plausibly lose control of such a situation. This risk is one that's present even with narrow AI, but grows as levels of AI intelligence and autonomy increase. So, if some businessmen apply AI to be business weapon to attack or steal their business competitors' business secret, e.g. contract document, employee performance report, profit report, even business secret document. Then, AI will be one business competitor weapon more than business assistant role in any business market. So, whether AI is office assistant or business competitor weapon, it depends on how the businessmen apply them to assist their business development.

The AI is programmed to do something beneficial, but it develops a destructive method for achieving its goal: This can happen whenever we fail to fully align the AI's goals with ours, which is strikingly difficult. If you ask an obedient intelligent car to take you to the airport as fast as possible, it might get you there chased by helicopters and covered in vomit, doing not what you wanted but literally what you asked for. If a superintelligent system is tasked with a ambitious geoengineering project, it might wreak havoc with our ecosystem as a side effect, and view human attempts to stop it as a threat to be met.

As these examples illustrate, the concern about advanced AI isn't malevolence but competence. A super-intelligent AI will be extremely good at accomplishing its goals, and if those goals aren't aligned with ours, we have a problem. You're probably not an evil ant-hater who steps on ants out of malice, but if you're in charge of a hydroelectric green energy project and there's an anthill in the region to be flooded, too bad for the ants. A key goal of AI safety research is to never place humanity in the position of those ants.

Because AI has the potential to become more intelligent than any human, we have no surefire way of predicting how it will behave. We can't use past technological developments as much of a basis because we've never created anything that has the ability to, wittingly or unwittingly, outsmart us. The best example of what we could face may be our own evolution. People now control the planet, not because we're the strongest, fastest or biggest, but because we're the smartest. If we're no longer the smartest, are we assured to remain in control?

A captivating conversation is taking place about the future of artificial intelligence and what it will/should mean for humanity. There are fascinating controversies where the world's leading experts disagree, such as: AI's future impact on the job market; if/when human-level AI will be developed; whether this will lead to an intelligence explosion; and whether this is something we should welcome or fear. But there are also many examples of of boring pseudo-controversies caused by people misunderstanding and talking past each other. To help ourselves focus on the interesting controversies and open questions — and not on the misunderstandings — let's clear up some of the most common myths.

There have been a number of surveys asking AI researchers how many years from now they think we'll have human-level AI with at least 50% probability. All these surveys have the same conclusion: the world's leading experts disagree, so we simply don't know. For example, in such a poll of the AI researchers at the 2015 Puerto Rico AI conference, the average (median) answer was by year 2045, but some researchers guessed hundreds of years or more. There's also a related myth that people who worry about AI think it's only a few years away. In fact, most people on record worrying about superhuman AI guess it's still at least decades away. But they argue that as long as we're not 100% sure that it won't happen this century, it's smart to start safety research now to prepare for the eventuality. Many of the safety problems associated with human-level AI are so hard that they may take decades to solve. So, any businessmen ought have business moralty to know whether they ought how to apply their AI to assist their business development in our future office environment to be more moral.

● Five ways to use AI to improve business efficiency to these office tasks

Regardless of a company's size or type, its executives typically look for ways to help it operate as efficiently as possible. They understand the link between efficiency and profitability. If employees waste too much time with drawn-out processes or complicated tasks, it'll be hard for the enterprise to remain profitable and adapt to challenges. Fortunately, artificial intelligence (AI) supports the need for effective business operations. Here are five ways enterprises can use AI for help: 5 ways to use AI to improve business efficiency image.Getting the best results from AI means looking at where bottlenecks exist, then figuring out if and how it might remove or minimise them. AI can help any offices to improve or raise efficiency to these tasks aspects as below:

1. Use AI to answer queries and support customer engagement

Chatbots are an increasingly popular option for businesses to try, and they use AI to work. Companies often build chatbots that can answer any questions from customers that come through outside of business hours. Some identify the nature of a person's problem, then either attempt to tackle it with preprogrammed answers or pass the communications to a human support worker. The retail industry, in particular, saw success by deploying chatbots. Global data collected by Juniper Research shows an estimated 2.6 billion retail-based chatbot interactions in 2019, and the company forecasts the number to rise to 22 billion in 2023.

Chatbots are excellent for answering simple questions like "How late are you open today?" or "Do you have gluten-free menu options?" Getting quick answers to queries like those increases the chances customers will choose to do business with one company over another. Equally importantly, when chatbots can give responses in a matter of seconds, there's no need for humans to stop what they're doing and address the questions.

2. To enhance reporting speed and accuracy

Company reports reveal things such as which products are selling the fastest and where they're most popular. They can also confirm the impacts of marketing campaigns on product sales, break down the costs of a new packaging choice or shipping method, and much more. However, as anyone that files reports knows, creating them is a painstaking task, and trying to rush through the process could cause mistakes. Some forward-thinking companies are combining AI with big data analytics. Doing this brings better forecasts and takes some of the burdens off the people who prepare the reports. AI also helps conquer the inevitability of mistakes. Even the most careful people make blunders, often because of mental fatigue.

AI learns to spot patterns in data and gets smarter with time. This means reports get finished faster and contain more-reliable information. The reliability aspect is crucial, especially since recently published research indicated two-thirds of the senior executives polled had no confidence or trust in big data. Using AI does not mean companies can do without data scientists. However, depending on the technology allows them to reduce the uncertainty that may otherwise exist. It also prevents employees who work with a company's data from being asked to recheck the findings, even if they initially took appropriate precautions to ensure accuracy.

3. To improve data transfer speeds

Fast data transfers help AI technology work. Concerning some information-intensive applications like virtual reality (VR), any slow transmissions greatly interfere with the realism, and content immersion people should enjoy after strapping on a VR headset. As it turns out, AI can improve data transfer speeds, too. For example, services exist that boost speeds across any wide-area network (WAN). Users enjoy consistently accelerated rates regardless of the kind of information transferred. Some companies have solutions that can reduce WAN job times by up to 98%. These AI-driven options work particularly well when companies need to move information between data centres or cloud environments.

4. To assist the IT team with identifying genuine cyberthreats and anomalies

One of the ongoing challenges faced by IT teams of all sizes is to separate the true cyber threats from false alarms. The difficulties associated with categorising the two types may mean cybersecurity professionals waste time getting to the bottom of things that are ultimately nonissues. They might miss the actual threats that could derail a company's operations. Besides detecting possible intrusions associated with a network, AI can screen for software abnormalities that may make it easier for cybercriminals to orchestrate their attacks successfully. It can also find malicious software hackers installed. Due to this kind of information and the advantages of receiving it through real-time updates, IT security teams can work more productively. They can use the majority of their resources on the threats that matter most to the company's stability.

Some organisations have even used AI to help them conquer the substantial skills shortage in the cybersecurity industry. At Texas A&M University, the Security Operations Center deals with about a million attempted hacks each month. The facility has some full-time workers, but students comprise most of the staff. They work alongside AI that aids in threat monitoring, detection and remediation. Before students see possible threats, the smart technology finds and groups them. This approach saves time and lets the team get to work investigating the problems and deciding how to handle them.

5. To streamline the time-to-hire metric when filling new positions

Statistics show the average time required to hire a person for an open position ranges from 12.7 to 49 days, depending on the industry. The timing also varies based on the type of work a job requires. For example, it takes a shorter amount of time overall to find someone for an administrative or human resources position than one associated with a creative or advertising role. Then, of course, interviews are more extensive for high-profile work.

Human resources professionals increasingly use AI to cut down on the time between first posting a job and finding the ideal individual to hire. For example, an AI platform could look for particular desired keywords in submitted resumes, saving hiring managers from poring over the documents themselves. AI can also pitch in during interviews. A company called VCV recently raised $1.7m to further develop its AI tool that has voice and facial recognition components. Candidates are asked to record videos of them answering interview questions, but they can't prepare for the specific content in advance.

In conclusion, AI Can Boost Efficiency at All Types of Companies. The examples here highlight why so many company leaders conclude that if they use AI, they could cut down on inefficiencies. Getting the best results from AI means looking at where bottlenecks exist, then figuring out if and how it might remove or minimise them. But, AI still lack enough effort to help all staffs to raise efficiency to all department tasks in any office environments.

- How the office energy Department is using AI to solve some of their office staffs electricity toughest challenges in their office working environment.

Insights from artifical intelligence has the potential to transform nearly every aspect of the world as we know it. Today, it is being applied to accelerate the pace of discovery in a wide variety of areas including energy, materials science, health care, national security, emergency response, transportation, and more. AI can be trained to help any energy department to gather data to avoid energy waste to be used to any organizations. So, AI is such as one super machine to do more accurate judgement to help any energy scientists to find the best methods to help any organizations to avoid to waste to use any energy daily. Then, organizations can avoid to spend too much energy to use in offices and they can save more money and avoid energy shortage challenge causes more easily. When the office managers can apply AI ability to reason and put it into a more automated format in a computer system to their every staffs' computer and record their computer electricity use record in their offices every day.

How can AI help offices to save energy or avoid to waste energy ?

The next industrial revolution is already happening. Artificial intelligence (AI) is ushering in an era of technologies that are faster, more adaptable, more efficient, and making the world more digitally connected. AI is best described as complementary to human intelligence, delivering the computing power to crunch numbers too big for people and recognize patterns too tedious for the human eye. In a Harvard Business Review study of 1,500 companies, it was found that the most significant performance improvements were made when humans and machines worked together. As AI becomes one of society's greatest assets, it's especially helpful for solving problems that seem larger than life — like protecting our natural environment.

Through machine learning, robotics, drones, and the internet of things (IoT), society can achieve better monitoring, understanding, and prevention of damage and stressors on Earth's land, air, and water. Even technology already available today could reduce energy usage in the U.S. by 12 to 22 percent, according to The Information Technology Industry Council (ITI). In the face of this dire reality, the potential of technology to help meet this challenge is a rare source of optimism. According to a recent survey by Intel and the research firm Concentrix, 74 percent of business-decision makers working in environmental sustainability agree artificial intelligence (AI) will help solve long-standing environmental challenges; 64 percent agree the Internet of Things (IoT) will help solve these challenges. As the field of AI develops, so will the potential to protect the environment. From the land and air to both drinking and ocean water, AI is shaping up to be the key that governments, organizations, and individuals can tap to work toward a cleaner planet, even AI can help offices to avoid to waste energy when staffs are working in offices every day.

Many AI scientists indicate that AI will also make renewable energy technology like solar panels and wind turbines more efficient and cost effective, helping them to become ubiquitous and lower society's dependence on fossil fuels. AI will also make renewable energy technology like solar panels and wind turbines more efficient and cost effective, helping them to become ubiquitous and lower society's dependence on the fossil fuels polluting the air — then hopefully eliminate them all together. Combined with the smart grid, another technology that will be enabled by AI, this will truly progress the way people receive and use electricity in their homes, offices, and everywhere else. Smart meters save energy by allowing for two-way communication between the grid and anything that uses electricity, giving energy providers a better understanding of usage and the ability to make real-time adjustments for efficiency. Customers will benefit from the real-time data too; seeing the increased costs at peak times will encourage them to voluntarily adjust their usage to save money. This will, in turn, save even more energy: a win-win. Plus, the process of delivering the energy itself will also be improved by the smart grid, thanks to Volt/VAR control systems that can reduce the amount of energy wasted when it's in electricity transmission lines.

Can AI replace office workers

Can AI replace all office workers to do their different tasks in office different department ? If AI can only replace some department office workers to do their simple tasks, how it can raise more efficiency to compare them in some business office environments. I shall indicate some office tasks to explain how AI can help these businesses to raise their efficiency in officesas below:

- AI insurance workers

Nowadays, some country offices begin apply robotics to replace human office workers in their companies. For example, Japanese company replaces office workers with artificial intelligence in insurance industry. A future in which human workers are replaced by machines is about to become a reality at an insurance firm in Japan, where more than 30 employees are being laid off and replaced with an artificial intelligence system that can calculate payouts to policyholders.

Fukoku Mutual Life Insurance believes it will increase productivity by 30% and see a return on its investment in less than two years. The firm said it would save about 140m yen (£1m) a year after the 200m yen (£1.4m) AI system is installed this month. Maintaining it will cost about 15m yen (£100k) a year. The move is unlikely to be welcomed, however, by 34 employees who will be made redundant by the end of March.

The system is based on IBM's Watson Explorer, which, according to the tech firm, possesses "cognitive technology that can think like a human", enabling it to "analyse and interpret all of your data, including unstructured text, images, audio and video".The technology will be able to read tens of thousands of medical certificates and factor in the length of hospital stays, medical histories and any surgical procedures before calculating payouts, according to the Mainichi Shimbun.

While the use of AI will drastically reduce the time needed to calculate Fukoku Mutual's payouts – which reportedly totalled 132,000 during the current financial year – the sums will not be paid until they have been approved by a member of staff, the newspaper said.

Japan's shrinking, ageing population, coupled with its prowess in robot technology, makes it a prime testing ground for AI. According to a 2015 report by the Nomura Research Institute, nearly half of all jobs in Japan could be performed by robots by 2035. For example, one Japan insurance company, Dai-Ichi Life Insurance has already introduced a Watson-based system to assess payments - although it has not cut staff numbers - and Japan Post Insurance is interested in introducing a similar setup, the Mainichi said. AI could soon be playing a role in the country's politics. Next month, the economy, trade and industry ministry will introduce AI on a trial basis to help civil servants draft answers for ministers during cabinet meetings and parliamentary sessions. The ministry hopes AI will help reduce the punishingly long hours bureaucrats spend preparing written answers for ministers.

- AI public service workers

The automated city: do we still need humans to run public services? If the experiment is a success, it could be adopted by other government agencies, according the Jiji news agency. If, for example a question is asked about energy-saving policies, the AI system will provide civil servants with the relevant data and a list of pertinent debating points based on past answers to similar questions.

The march of Japan's AI robots hasn't been entirely glitch-free, however. At the end of last year a team of researchers abandoned an attempt to develop a robot intelligent enough to pass the entrance exam for the prestigious Tokyo University. "AI is not good at answering the type of questions that require an ability to grasp meanings across a broad spectrum," Noriko Arai, a professor at the National Institute of Informatics, told Kyodo news agency. Hence, AI will have possible to replace some public service workers' tasks.

- AI replace warehouse workers

Denso's use of Drishti shows how some jobs will be transformed by artificial intelligence even when they're unlikely to be eliminated by AI anytime soon. Many jobs in manufacturing require dexterity and resourcefulness, for example, in ways that robots and software still can't match. But advances in AI and sensors are providing new ways to digitize manual labor. That gives managers new insights—and potentially leverage—on workers. For example,some workers say the results are unpleasant. Last year, Amazon warehouse employees in Minnesota staged a walkout to protest how the company uses inventory and worker-tracking technology. They allege that Amazon uses it to enforce a punishing working pace that causes injuries. The company has disputed those claims, saying it coaches employees on how to safely meet quotas.

Workers at Denso were initially wary of the prospect of being video-recorded all day to feed machine-learning algorithms, but Huffman says they have since come to appreciate Drishti's technology. After something goes wrong, workers can now look at the data and video with their managers, instead of having to hope bosses take their account of what happened seriously. Huffman says having a constant readout on productivity also helps managers be more responsive to nascent problems. "If somebody's struggling, not every associate is going to call for help," he says. "If we see their cycle time is jumping through the roof, we can go over and say 'Are you having any issues?'"Workers on Denso lines equipped with Drishti's technology now get a personal feed of their own data. Monitors on each workstation display how a worker is doing, says Raja Shembekar, a Denso vice president. If the worker completes their assembly step on time, they see a smiley face—if not, a frowny one. Hence, Amazon had begun to apply AI robotic to replace some warehouse workers' tasks.

For another factory manufacture working environment example, AI can replace many manufacture workers to do their tasks in factories. Route 9 skims by Boston and cuts clear across Massachusetts to Pittsfield, a city of roughly 50,000, the largest in Berkshire County. Well east of Pittsfield, Route 9 becomes Worcester Road, named for a city that in earlier times was the nation's largest manufacturer of wire—barbed wire, electrical wire, telephone wire and the wire used in the making of undergarments by the Royal Worcester Corset Co., once the largest employer of women in the United States. Older Worcester residents can still recall the factory bells pealing to signal the start and end of the workday. Now, the bells are silent, and the wire and corset factories have been replaced with three of the nation's largest employers: Walmart, Target and Home Depot. If this sounds familiar, it should. It has been nearly two decades since retail overtook manufacturing as the nation's most important job creator, employing roughly one of every 10 American workers—more people than in health care and construction combined. That's a lot of jobs.

Of course, not all retail jobs qualify as what most of us consider good jobs. Today, the average hourly wage for a nonsupervisory retail worker is $11.24, and less than half of retail workers receive benefits of any kind. Still, as a nation, we've come to a sort of uneasy peace with this trend. We know that manufacturing employs far fewer Americans today than it once did—that iPads and Macs aren't made in America and neither are many televisions, appliances, tools, toys or clothes. We also know that shopping for these appliances, tools, toys and clothes is an all-American pastime: On average, we spend nearly 45 minutes a day (more than 270 hours per year) purchasing goods and services. Retail has become the world as we know it, and many of us expect to make our living working in that world.Thanks to automation and a killer business model, Amazon is so efficient that it reaps nearly twice the revenue per employee of Walmart, despite the fact that Walmart, too, has a substantial online presence. Worldwide, Amazon has installed over 100,000 robots to labor in "perfect symbiosis" with humans in its warehouses and has plans to install many thousands more. While it's not clear what constitutes perfect symbiosis, the robots are said to save the company $22 million annually, per warehouse. The company's master plan of an autonomous future also includes goods delivered by drones and self-driving vehicles.

For while Amazon continues to open warehouses around the globe and staff them with many thousands of human beings, estimates are that every human on the Amazon payroll—whether full- or part-time—displaces two humans at traditional brick-and-mortar operations. And that's a feature, not a bug: As Tim Lindner, a veteran IT analyst, confided in a note to industry insiders, eradicating jobs is the explicit goal of any online retailer. As he once wrote: "Labor is the highest-cost factor in warehouse operations. It is no secret that Amazon is moving to highly automated operations within its distribution centers, and...it has additional technology that can further reduce the number of humans it needs to process customer orders.... You have heard the old programmer's phrase, 'Garbage in, garbage out.'... [With] the diminishing reading abilities of humans on the Receiving dock, finding an automated solution to eliminate the 'garbage in' problem is the holy grail. Amazon may have just patented it."

By garbage, Lindner meant human error, the alternative to which is apparently robotic precision. And robots can be very precise, especially when it comes to routine tasks. Sawyer, an industrial robot created by the former Boston-based Rethink Robotics, offers an impressive illustration of how all-embracing a robot arm can be. Sawyer is the brainchild of Rodney Brooks, the inventor of both Roomba, the robotic vacuum, and PackBot, the robot used to clear bunkers in Iraq and Afghanistan and at the World Trade Center after 9/11. Unlike Roomba and PackBot, Sawyer looks almost human—it has an animated flat-screen face and wheels where its legs should be. Simply grabbing and adjusting its monkey-like arm and guiding it through a series of motions "teaches" Sawyer whatever repeatable procedure one needs it to get done. The robot can sense and manipulate objects almost as quickly and as fluidly as a human and demands very little in return: While traditional industrial robots require costly engineers and programmers to write and debug their code, a high school dropout can learn to program Sawyer in less than five minutes. Brooks once estimated that, all told, Sawyer (and his older brother, the two-armed Baxter robot) would work for a "wage" equivalent of less than $4 an hour.

Robots loom large in discussions of work and its future, a conversation that can get mired in false assumptions. Until recently, many economists were skeptical that automation could permanently displace human workers on a large scale. People have always shifted away from work better done by machines, but the economic principle of "comparative advantage" predicts that humans will maintain an edge in many fields. Under this logic, technology will not displace us but set us free to do less dangerous, more challenging things, essentially the very things that make humans human. Of course, human workers are complicated. We get tired, hungry, distracted, angry, confused. We make mistakes, sometimes egregious ones. Machines lack our frailties and biases and are better equipped to weigh evidence fairly, without prejudice or false assumptions. Perhaps most critically, machines can retain and process data far more accurately than we can, and that data is growing exponentially.

Every minute of every day, Google services 3.6 million searches in the United States alone. Spammers send 100 million emails. Snapchatters send 527,000 photos, and the Weather Channel broadcasts 18 million forecasts. This and more data—properly collected, codified and analyzed—can be applied to automate almost any high-order task. Data can also serve as a surrogate for human experience and intuition. Online shopping and social media sites "learn" our preferences and use that information to make values-based assessments to influence our decisions and behavior. And, increasingly, machines excel in the tasks once thought uniquely human."Computers are able to see and hear, and have face-recognition capabilities that are significantly better than humans," says Vardi. "Machines understand the human world far better than they did just a few years ago. And we haven't discovered anything in the human brain that can't be modeled."

● AI can replace counter cashier service staffs

And robots need not be perfect, only equal to—or a tad better than—complicated and expensive humans. And technologists are working hard to make sure they are a tad better. For example, in the case of retail, it's become clear that many of us avoid the self-service checkout line—we prefer the cashier to punch in our purchases rather than do so ourselves. So it seems that the job of cashier—among the largest retail employment categories—is not directly at risk. But Zeynep Ton, an MIT management expert who focuses on the retail sector, says self-service checkout is only a first step and not a terribly smart one. "Customers recognized that self-service checkout is not an innovation, but merely a way of outsourcing the job to them, so they didn't like it," she says. "But new technology is coming that will make self-service checkout so much easier and faster, and that will have a real impact on retail employment."

Experts caution that the so-called apocalypse in retail predicted a few years ago has not yet come to pass. In fact, for every company closing existing stores, two more are opening new stores. Retail is a highly competitive industry, and technology is transforming not only the way we shop but the way we connect with brands—for example, just a few years ago, who would have imagined that Amazon would open actual retail stores? And while e-commerce has grown to 10 percent of retail, that still leaves 90 percent for brick-and-mortar stores. But those brick-and-mortar stores, too, are undergoing radical change that has serious implications for America's workforce.

As example, Lobaugh cites food trucks, which he says increasingly pose a threat to many fast-food outlets. Unlike restaurants pinned down by a pair of Golden Arches, food trucks are nimble—they can home in on areas where customers are most likely to gather at any particular time. They can also tailor their offerings to a particular region or even a neighborhood, as well as use Facebook or other media to get out the word on their menu items and locations. Small, specialty stores also have far more flexibility than large department stores. "Technology has reduced the cost of entry into new markets, so in retail there are fewer big, monolithic companies, but more small competitors," he says. "Companies are diversifying to meet the specific needs and desires of consumers—everyone's piece is getting smaller, but there are many more pieces."

But despite what it predicts will be a banner holiday season, this year Amazon took on far fewer seasonal employees than usual—100,000 employees versus 120,000 the previous two years. And while an Amazon spokeswoman insisted that automation is not a factor in this reduced workforce, others seem to not agree. In a recent report, Morgan Stanley analyst Brian Nowak soothed the fears of Amazon shareholders concerned with the wage increase by pointing out that automation had already and would continue to reduce the call for labor, and therefore reduce overall costs. When asked about this, Lobaugh again tactfully declined to comment—other than to say that while the retail sector had lost less ground than most people assume, retail employees were another matter. "There are winners," he says, "and then there are losers."

- AI can replace accountants in accountancy service industry

Not that long ago artificial intelligence (AI), robots and machine learning (ML) were thought to be things only found in science fiction films. Today, this type of technology is taking center stage in workplaces across the globe. Industries, including manufacturing, retail, agriculture, and customer service have already had AI replace some job positions that left workers scrambling to find new career options. This AI revolution is not expected to slow down anytime soon. In fact, experts anticipate that as many as 800 million jobs could be replaced with AI technology by the year 2030. Initially, AI technology and automation in the workplace seemed to only affect pink and blue-collar workers. As this technology advances and becomes more powerful, professional, white-collar workers, including accountants, are starting to worry about what the future holds for their career and if AI will be developed to own their professional skills in accounting service industry.

In basic terms, AI technology is intelligent machines that are able to complete repetitive, mundane tasks at a fraction of the time it takes humans and with greater accuracy. The emergence of Machine Learning now allows AI platforms to observe, analyze and self-learn data and processes to improve its performance and accuracy over time. AI technology is already able to handle many accounting functions, such as tax preparation, payroll, and audits. Many of the leading accounting software providers, including Xero, Intuit and Sage have incorporated AI technology into their software to handle basic accounting tasks, such as bank reconciliations, invoice categorization, risk assessment, and audit processes, like expense submissions and invoice payments. Many of these standard tasks are extremely time-consuming, which has many accountants across the country worried about how the emerging AI technology will affect their billable hours. An even bigger concern is that AI technologies will replace the need for companies to work with accountants at all.

- AI Will Transform not Replace Accountants

While there is no doubt that AI technology is capable of handling many standard accounting tasks faster and more efficiently or that these capabilities will only increase over time, it doesn't mean the end for accountants. There always will be a need for that human element - human intelligence - at the other end of AI technology. In fact, according to leading research firm, Gartner, AI is set to create more jobs than it will replace, leaving workers, including accountants with options. Accountants don't have to worry about their job being replaced by AI any time in the near future. Companies will always need accountants that can analyze and interpret AI data, as well as provide consulting services. Rather than replacing the role of an accountant, AI technology will transform the duties an accountant performs.

With AI technology and machine learning handling many of the mundane, repetitive tasks, accountants will have more time to focus on other aspects of the job, such as consulting and data analysis. This is good news for many accountants. Rather than spending hours completing menial tasks, accountants of the future will be able to use and analyze AI data to provide their clients with sound business solutions.

In many ways, AI will help accountants improve their services. AI technology will improve data entry accuracy and lower the liability risk for accountants. In addition, emerging technology is more efficient at fraud detection, adding an extra layer of protection for accountants and their clients. It also provides real-time data, which allows accountants to provide real-time solutions. Even more impressive is the ability of machine learning to analyze large amounts of data instantly, evaluate past successes and failures in an effort to accurately predict future outcomes.

There is no way to escape the use of AI technology, at least not if you hope to remain competitive in the upcoming years. The speed, efficiency and accuracy of AI technology just cannot be beat. The only thing accountants can do is to embrace this new technology and learn how to maximize its use. The better equipped you are to help your clients integrate and utilize AI technology in their accounting processes the more valuable you will be. For example, many universities today are already incorporating IT and database management courses into their accounting program. This means that graduating students are coming into the workforce with the skills they need for future accounting work. Accountants already in the workforce must find ways to acquire these skills in order to remain relevant to their employers and/or their clients. Accountants can obtain the IT skills they need by attending seminars, using self-learning online programs or attending college-level courses. It is equally important for accountants to stay up-to-date on the latest accounting trends, emerging technologies and industry news. This will allows accountants to not only keep their jobs but to also provide more efficient services to their clients. Rather than worry about AI taking over their jobs, accountants should embrace this technology as a powerful solution to enhance customer services. Finally, accountants will be able to use all their training and experience to provide customer will real and effective business solutions, whether it's in reference to tax consulting, real estate deals, mergers, growth options, or any other business practice.

On conclusion, technology is advancing at record rates so now is the time to obtain the IT and database management skills you need to advance into the future. With the right skills and training, accountants are guaranteed a lucrative career that will last well into the future.

Why Developed And Developing Countries Need Artificial Intelligent Development To Assist Office Tasks

Must developed and developing countries need artificial intelligent development to assist office tasks ? Ought AI is needed to prefer to develop technique to assist office staffs to reduce workload to compare other kinds of occupation environment tasks aspects ? If one developed country, e.g. US, UK , Japan , Singapore it does not continue to develop artificial intelligence, robotic, then what disadvantges or weaknesses , it will encounter to compare when it chooses to continue to develop this artificial intelligent technology in society. If one developing country, e.g. China, Korea, Taiwan, it does not continue to develop artificial intelligence, robotic, then what disadantages or weaknesses, it will also encounter to compare when it chooses to continue to develop this artificial intelligent technology in in society. I shall explan the reasons why the results may cause to either the developed country, or the developing country as below:

- How AI help developing countries to communication and agriculture and learning and medical delivery development

Why can AI help developing countries ? Drones that pick inaccessible crops and mobile phones that give medical advice are two of the ways AI can transform life in the developing world. Artificial intelligence (AI) may improve the lives of the world's poor, the technology needed to revolutionise inefficient, ineffective food and healthcare systems in developing countries is well. For example, in low-income areas, agriculture and healthcare are two critical ecosystems that we can apply AI to immediately; this is not the far future, or even in five years.

Artificial intelligence (AI) has seeped into the daily lives of people in the developed world. From virtual assistants to recommendation engines, AI is in the news, our homes and offices. There is a lot of potential in terms of AI usage, especially in humanitarian areas. The impact could have a multiplier effect in developing countries, where resources are limited.

Emergency Response to developing countries' earthquake natural damage suddence occurrence predicting

AI and machine learning are still finding importance in emerging markets, but certain applications have emerged and are now widely used. For instance, predictive models for disaster relief enable first responders to automatically analyze large-scale behavior and movement through multiple sources of data including social media platforms, web forums, news sources, etc. Based on collected data, responders can scale reconstruction efforts and distribute supplies in a timely manner.

Why and how AI can assist farmers to predict when the earthquake occurs suddenly in order to avoid or reduce the natural damage to their agriculture productive number loss. For example, In 2015, when a major earthquake hit Nepal, more than 8 million people were affected. During the aftermath, drones were used to map and assess the destruction and speed up the rescue mission. The town of Sankhu, situated about 20 kilometers northeast of Kathmandu, was among the highly affected locations. In May 2018, my company Fusemachines and GeoSpatial Systems partnered with Sankhu's city officials to use drones and artificial intelligence in an effort to automatically estimate the reconstruction need. After processing data accumulated from a drone-powered aerial mapping of the region, the team fed this data to advanced machine learning algorithms. Combining drone imagery, digital mapping and machine learning, the team configured region modeling and infrastructure development with higher accuracy. Another organization known as One Concern, a California-based startup, has created a predictive AI program called Seismic Concern to accurately predict seism and is also working on solutions for wildfires, floods and hurricanes.

Smart AI Agriculture

Another application of AI in developing countries is smart agriculture. Farmers monitor crops more effectively and make better predictions on planting, weeding and harvesting using AI tools. It can also be used to analyze one plant at a time and add pesticides only to infected plants and trees instead of spraying pesticides across large swaths of crops. One California-based tech company is an example of this use of AI. So, the developing countries farmers in rural parts of India are also using AI to increase yields through better access to information about the farming season than they would normally have. Technology-enabled process automation offers the agribusiness industry the chance for remarkable growth -- not only in developed countries but around the world. There's a unique opportunity to increase yields, cut down labor costs and improve people's health.

Medicine Delivery to developing countries' patients urgent need

Companies are also leveraging AI to improve access to health care in some of the most remote areas of the world. In Rwanda, for example, Zipline is using drones to deliver medical supplies and blood to hospitals and clinics that are difficult to access by car. This has dramatically impacted people living in remote parts of the country because they are able to get medical help when needed. The drone system in Rwanda has also helped reduce waste of blood by 95%, as noted by Zipline. One Concern has created an AI program called Seismic Concern that accurately predicts seismic events and is also working on solutions for floods, wildfires and hurricanes. The medical field may actually benefit the most from emerging technologies in developing countries.

Assistance to reduce teaching work workload or psychological pressure to teachers in developing countries' schools

Another vital area benefiting from innovative technologies like AI is education. Advanced technologies can enhance how we learn, teach and perform tasks. In most developing countries, schools lack experienced teachers and resources to enhance students' knowledge. As a result, many students still have to walk long distances to get to the nearest school, which has created education gaps, especially in rural areas. AI tools such as personalized learning assistants can simplify learning by making tutoring services and learning materials accessible to all students, wherever they are. Machines can be automated to help students learn basic concepts without a tutor, which companies like Carnegie Learning are working on. This would allow students to learn at any time from anywhere. With AI, education is made easy and accessible to more people.

The initial usage of AI in developing countries has been at a micro level -- solving small, specific problems in a defined industry. As machine learning advances and there is a higher utilization of AI, we will see more complex issues being targeted and resolved. When duly adopted, AI can positively impact future developing countries people everyday lives not just in disaster intervention, education, health care and agriculture but can also help in mitigating poverty, malnutrition and pollution. Especially, in developing nations, to leverage AI's true potential and create a snowball effect. Startups are defining a holistic and humanitarian approach to building more sophisticated, AI-ready societies. Stakeholders in the AI landscape should understand the strengths and nuances of the developing world as well as the limitations of AI and create localized solutions and applications.

Why does smart phone help developing countries communication ?

Internet Seen as Positive Influence on Education but Negative on Morality in Emerging and Developing Nations. Internet access differs substantially across the 32 emerging and developing countries polled, with the lowest rates of internet use in South Asian and sub-Saharan African nations. Within countries, computer owners, young people, the well-educated, the wealthy and those with English language ability are much more likely to access the internet than their counterparts. To access the internet, people increasingly use smartphones rather than more cumbersome fixed landline connections and computers. Around the world, both smartphones and basic-feature phones alike are used for sending messages and taking pictures.

In fact, many developing countries young people, students are popular to use smart phones for internet usage aim, instead of communication. Moreover, many developing countries working people are also popular to use smart phones for any working usage in their working time , even non working time any time. So, smart phones (AI) phones will be important communication or leisure tools to developing countries people in the future. Unless, it is one day, scientists can develop another new communication tool to replace smart phones. So, artificial intelligence will be important to influence developing countries people , how to improve or bring positive learning attitudes to students in their daily learnnng lifes. as well as how to raise developing countries people, how to raise working people efficiency or improve performace in their daily working lifes. So, AI may bring positive learning or working attitudes to developing countries working people and students both.

The Positive Impact of Mass Media in Developing Countries

Radio, newspapers, television, Internet, social media, etc., all of these are forms of mass media. Each of these outlets has the capability of bringing information to thousands of people with one device. While in some communities it is easy to take advantage of these communication outlets such as television and Internet access, not everyone has access to such outlets. Radio is one of the most common forms of mass media in developing countries because it's affordable and uses less electricity than many other forms of mass media, but only approximately 75 percent of people in developing countries have access to a radio, and roughly 77 percent of people in rural areas have access to electricity.

For developing countries that have implemented forms of mass media in their communities, there have been numerous positive outcomes are influenced to impact developing countries mass media by artificial intelligence as below:

When AI is participated to developing countries mass media, it can influence any radio, television audiences raise more attention to each other through social media platforms such as Facebook and Twitter and create, organize and initiate street protests and campaigns. Furthermore, having access to social media in developing countries, people are able to connect to those that they usually wouldn't have the chance to talk to. Moreover, AI Provides educational opportunities- In many countries, the division between local and national languages as well as issues of literacy can make communication difficult. With the use of mass media, a bridge can be built between these two gaps. In India, there is a radio station that provides information in local languages and respects local culture and traditions. One of the main ways is to create public awareness of what is going on with businesses and government officials. The media plays an important role in giving people the opportunity to act against injustice, oppression and misdeeds that they otherwise wouldn't know about. Information on available healthcare, a mass radio broadcast was sent out encouraging parents to seek treatment at local healthcare facilities for their sick children. With this mass outreach on healthcare, the encouragement of people to take their children to healthcare facilities saved thousands of lives. This easy way of encouraging others and bringing awareness about certain diseases was made possible through a simple radio broadcast. Finally, when AI is particiapted to media, it may bring many social issues to life that otherwise would remain unknown to many people. In developing countries and communities like Burkina Faso, when the radio broadcast was released about malaria, diarrhea and pneumonia, people were educated and moved to action and knew to take their children to healthcare facilities for preventative care. As it is seen, having access to different media outlets is vital for those in developing countries. Here are three ways that those in developing countries can implement mass media to help their people and communities.

When AI is participated to any internet radio or internet newspaper mass online listening or reading channel. It can provide online radios or newspapers in public places- By providing online radios and newspapers in public areas it gives community members to access news, information and emergency warnings. Even though radios can be on the cheaper side, there are still many people that can't afford to have a radio in their home. By providing one in a local place, not only would it better educate the community members but also it will bring the community together. So, it can make media outlets a two-way platform- Creating a two-way platform between the community and those who are behind the radio stations, newspapers or broadcasts makes the community feel involved and that their voices are being heard. An organization called Soul City in sub-Saharan Africa is showing how well two-way platforms work by engaging their listeners and having them contribute thoughts and ideas about complex issues. Because developing countries radio listening audiences or newspaper readers are popular to accept computer online radio listening channel or online newspaper reading channel to replace traditional paper newspapers or radio machines. So, AI may raise their listening news or reading news leisure feeling from online mass media channel in the future.

● Why do developed countries need to develop AI

Artificial intelligence, or AI, is driving massive shifts across the globe, and every day more questions arise. What impact will AI have on the workforce and how can we prepare for it? How can we encourage economy-boosting and job-creating technologies? How can we ensure that AI will be implemented ethically and with minimal bias? How will society benefit? For developed country, such as US example. None of the US, Israel and Russia have a formal national AI policy yet. Private sector companies such as Google, Amazon and Apple and the US department of defence are driving the bulk of AI investment in the United States. Though Israel does not have a specific policy, it is keenly focused on AI and has seen the number of AI start-ups triple since 2014.

Developed country may learn whether what weakness it is lacking when it does not continue to develop AI from one another developed country. Which countries are approaching AI most effectively, and to what degree is there opportunity for greater international collaboration? It may be too early to tell; however, when analyzing the best practices of existing national AI policies, there is much that can be learned. These are the specific areas to consider. When one developed country continue to develop or research AI, it may bring these benefits as below:

On gathering Data aspect, from self-driving vehicles to smart cities, data is the driver behind AI. Innovation in the United States is limited without a national strategy that answers questions about protocol and ownership. France and Denmark, on the other hand, are opening government data. France is hosting troves of centrally collected public and private data that it plans to make available as part of its strategy. Conversely, by taking a restrictive position on issues of data collection (as indicated by the implementation of General Data Protection Regulation), the EU is putting manufacturers and software designers at a disadvantage while balancing the demand for privacy. On raising technologica talent aspect, the demand for AI talent far outweighs the available supply. As a result, almost every nation's strategy addresses talent development. Canada's AI strategy is distinct in that it primarily focuses on research and talent strategy. The country boasts AI degree programmes and is building a $127 million research facility in Toronto. Companies like Facebook and my own company, Uptake, are investing in Canada to access this talent pool. On AI legal technological innovation aspect, a whole host of legal questions swirl around AI. The country is developing a bill for AI liability that will be ready in March 2019. The government hopes the legal framework will attract investors by providing a simple, comprehensive guideline to enable the broad use of AI systems. So, when the developed country applied AI technology to assist any lawyers to work, then AI can help them to reduce the workload to draft any legal documents more easier. So, any developed countries lawyers' draft legal documents time must reduce if the developed countries lawyers accept to apply AI to assist their legal works. One of the great promises of AI is its potential for improving quality of life. But without the right planning and oversight, we risk exacerbating problems of inequality or marginalizing groups of people. As an example, India's AI strategy is focused on leveraging the technology not only for economic growth, but also for social inclusion.

AI may bring what benefits to developed countries

From SIRI to self-driving cars, artificial intelligence (AI) is progressing rapidly. While science fiction often portrays AI as robots with human-like characteristics, AI can encompass anything from Google's search algorithms to IBM's Watson to autonomous weapons. Artificial intelligence today is properly known as narrow AI (or weak AI), in that it is designed to perform a narrow task (e.g. only facial recognition or only internet searches or only driving a car). However, the long-term goal of many researchers is to create general AI (AGI or strong AI). While narrow AI may outperform humans at whatever its specific task is, like playing chess or solving equations, AGI would outperform humans at nearly every cognitive task.

Why research AI safety? Would AI bring war when AI is continued to develop by developed countries? In the near term, the goal of keeping AI's impact on society beneficial motivates research in many areas, from economics and law to technical topics such as verification, validity, security and control. Whereas it may be little more than a minor nuisance if your laptop crashes or gets hacked, it becomes all the more important that an AI system does what you want it to do if it controls your car, your airplane, your pacemaker, your automated trading system or your power grid. Another short-term challenge is preventing a devastating arms race in lethal autonomous weapons.

In the long term, an important question is what will happen if the quest for strong AI succeeds and an AI system becomes better than humans at all cognitive tasks. As pointed out by I.J. Good in 1965, designing smarter AI systems is itself a cognitive task. Such a system could potentially undergo recursive self-improvement, triggering an intelligence explosion leaving human intellect far behind. By inventing revolutionary new technologies, such a superintelligence might help us eradicate war, disease, and poverty, and so the creation of strong AI might be the biggest event in human history. Some experts have expressed concern, though, that it might also be the last, unless we learn to align the goals of the AI with ours before it becomes superintelligent.

There are some who question whether strong AI will ever be

achieved, and others who insist that the creation of superintelligent AI is guaranteed to be beneficial. At FLI we recognize both of these possibilities, but also recognize the potential for an artificial intelligence system to intentionally or unintentionally cause great harm. We believe research today will help us better prepare for and prevent such potentially negative consequences in the future, thus enjoying the benefits of AI while avoiding pitfalls.

How can AI be dangerous when developed countries continue to develop AI to become weapon to replace soldiers?

Most researchers agree that a superintelligent AI is unlikely to exhibit human emotions like love or hate, and that there is no reason to expect AI to become intentionally benevolent or malevolent. Instead, when considering how AI might become a risk, experts think two scenarios most likely:

The AI is programmed to do something devastating: Autonomous weapons are artificial intelligence systems that are programmed to kill. In the hands of the wrong person, these weapons could easily cause mass casualties. Moreover, an AI arms race could inadvertently lead to an AI war that also results in mass casualties. To avoid being thwarted by the enemy, these weapons would be designed to be extremely difficult to simply "turn off," so humans could plausibly lose control of such a situation. This risk is one that's present even with narrow AI, but grows as levels of AI intelligence and autonomy increase.

The AI is programmed to do something beneficial, but it develops a destructive method for achieving its goal: This can happen whenever we fail to fully align the AI's goals with ours, which is strikingly difficult. If you ask an obedient intelligent car to take you to the airport as fast as possible, it might get you there chased by helicopters and covered in vomit, doing not what you wanted but literally what you asked for. If a superintelligent system is tasked with a ambitious geoengineering project, it might wreak havoc with our ecosystem as a side effect, and view human attempts to stop it as a threat to be met. So, a super-intelligent AI will be extremely good at accomplishing its goals, and if those goals aren't aligned with ours, we have a problem. You're probably not an evil ant-hater who steps on ants out of malice, but if you're in charge of a hydroelectric green energy project and there's an anthill in the region to be flooded, too bad for the ants. A key goal of AI safety research is to never place humanity in the position of those ants.

Why the recent interest in AI safety ?

Stephen Hawking, Elon Musk, Steve Wozniak, Bill Gates, and many other big names in science and technology have recently expressed concern in the media and via open letters about the risks posed by AI, joined by many leading AI researchers. The idea that the quest for strong AI would ultimately succeed was long thought of as science fiction, centuries or more away. However, thanks to recent breakthroughs, many AI milestones, which experts viewed as decades away merely five years ago, have now been reached, making many experts take seriously the possibility of superintelligence in our lifetime. While some experts still guess that human-level AI is centuries away, most AI researches at the 2015 Puerto Rico Conference guessed that it would happen before 2060. Since it may take decades to complete the required safety research, it is prudent to start it now.

Because AI has the potential to become more intelligent than any human, we have no surprise way of predicting how it will behave. We can't use past technological developments as much of a basis because we've never created anything that has the ability to, wittingly or unwittingly, outsmart us. The best example of what we could face may be our own evolution. People now control the planet, not because we're the strongest, fastest or biggest, but because we're the smartest. If we're no longer the smartest, are we assured to remain in control?

A captivating conversation is taking place about the future of artificial intelligence and what it will/should mean for humanity. There are fascinating controversies where the world's leading experts disagree, such as: AI's future impact on the job market; if/when human-level AI will be developed; whether this will lead to an intelligence explosion; and whether this is something we should welcome or fear. But there are also many examples of of boring pseudo-controversies caused by people misunderstanding and talking past each other. When one developed country continue to develop AI, can itself country's all factories workers will lose jobs, due to AI can replace them to do simple works in factories, or any public transport drivers, e.g. bus drivers, ferry , tram, train drivers, they will lose jobs, when AI (non manual driving drivers) can replace all public transport drivers. So, some occupations will lose if developed countries continue to develop or research AI to replace human to do some simple jobs, such as some cooking jobs can be done by AI. So, it is possible that future cookers won't be needed, because AI cooking skills may be better than them to cook any good taste chinese or western food in restaurants. If you drive down the road, you have a subjective experience of colors, sounds, etc. But does a self-driving car have a subjective experience? Does it feel like anything at all to be a self-driving car? Although this mystery of consciousness is interesting in its own right, it's irrelevant to AI risk. If you get struck by a driverless car, it makes no difference to you whether it subjectively feels conscious. In the same way, what will affect us humans is what superintelligent AI does, not how it subjectively feels.

In fact, AI may be make any brokers jobs in financial market. the main concern of the beneficial-AI movement isn't with robots but with intelligence itself: specifically, intelligence whose goals are misaligned with ours. To cause us trouble, such misaligned superhuman intelligence needs no robotic body, merely an internet connection – this may enable outsmarting financial markets, out-inventing human researchers, out-manipulating human leaders, and developing weapons we cannot even understand. Even if building robots were physically impossible, a super-intelligent and super-wealthy AI could easily pay or manipulate many humans to unwittingly do its bidding. So, future brokers will be replaced by AI, when AI can be made to own financial brokers' analytical mind to make more accurate whether the share price will rise up or fall down to compare human financial brokers' analytical mind. The robot misconception is related to the myth that machines can't control humans. Intelligence enables control: humans control tigers not because we are stronger, but because we are smarter. This means that if we cede our position as smartest on our planet, it's possible that we might also cede control.

Not wasting time on the above-mentioned misconceptions lets us focus on true and interesting controversies where even the experts disagree. What sort of future do you want? Should we develop lethal autonomous weapons? What would you like to happen with job automation? What career advice would you give today's kids? Do you prefer new jobs replacing the old ones, or a jobless society where everyone enjoys a life of leisure and machine-produced wealth? Further down the road, would you like us to create superintelligent life and spread it through our cosmos? Will we control intelligent machines or will they control us? Will intelligent machines replace us, coexist with us, or merge with us? What will it mean to be human in the age of artificial intelligence?

Why do developed countries people need AI ?

Why do we assume that AI will require more and more physical space and more power when human intelligence continuously manages to miniaturize and reduce power consumption of its devices. How low the power needs and how small will the machines be by the time quantum computing becomes reality? Why do we assume that AI will exist as independent machines? If so, and the AI is able to improve its Intelligence by reprogramming itself, will machines driven by slower processors feel threatened, not by mere stupid humans, but by machines with faster processors? What would drive machines to reproduce themselves when there is no biological incentive, pressure or need to do so?

Who says superior AI will need or want to have a physical existence when an immaterial AI could evolve and preserve itself better from external dangers. What will happen if AI developed by competing ideologies, liberalism vs communism, reach maturity at the same time, will they fight for hegemony by trying to destroy each other physically and/or virtually. If AI is programmed to believe in God, and competing AI emerges programmed by muslims, christians or jews, how are the different AI's going to make sense of the different religious beliefs, are we going to have AI religious wars? What if the "powers that be" greatest fear is the emergence of a super AI that police's and rationalizes the distribution of wealth and food. A friendly super AI that is programmed to help humanity by, enforcing the declaration of Human Rights (the US is the only industrialized country that to this day has not signed this declaration) ending corruption and racism and protecting the environment.Most benefits of civilization stem from intelligence, so how can we enhance these benefits with artificial intelligence without being replaced on the job market and perhaps altogether?

Key to the process of machine learning are neural networks. These are brain-inspired networks of interconnected layers of algorithms, called neurons, that feed data into each other, and which can be trained to carry out specific tasks by modifying the importance attributed to input data as it passes between the layers. During training of these neural networks, the weights attached to different inputs will continue to be varied until the output from the neural network is very close to what is desired, at which point the network will have 'learned' how to carry out a particular task. A subset of machine learning is deep learning, where neural networks are expanded into sprawling networks with a huge number of layers that are trained using massive amounts of data. It is these deep neural networks that have fuelled the current leap forward in the ability of computers to carry out task like speech recognition and computer vision.

In conclusion, when developed countries continue to develop AI, it may bring positive advantages to bring raising productivies, or efficiencies, but it may also raise unemployment ratio to any low skill or low knowledge jobs in ther societies. However, human future society will need to change to be better to raise our living standard. But AI is one kind the best choice tool to achieve this aim in our future, so I agree developed countries continue to develop or research AI to be the super -human machine.

Artificial Intelligence Worker Brings
Working Environment Influences

Robots were once known only for the manufacturing business but today they are very much part of many workplaces. The future is even more promising for this wonder of artificial intelligence.Imagine a robot doing some of the major tasks of managers like using data to evaluate problems, making better decisions, monitoring team performance, and even setting goals.

Technology is playing a pivotal role in helping humans work more effectively. Since automation has become an integral part of business operations, we can predict that robots are soon going to replace many jobs that are today performed by humans. Now that the corporate world is also on the cusp of entering the robotic age, let's see what pros and cons this technology offers business world. If one day, our global working environments have any kinds of robotic participates to our service and warehouse and office etc. different working environment in order to assist office workers, service workers, warehouse workers, professional lawyers, doctors accountants job duties, what positive or negative influences, it will bring to what negative or positive effects to any office , warehouse, shopping centre, hospital, transport , restaurant etc, different working environments. Can robotic help office , warehouse to raise efficency ? Can robotic help hospital, restaurant, cinema, shopping center to improve service performance? Can robotic influence working environment to be worse? Can robotic help office or any working places to reduce expenditure or reduce long time machine and salary cost when they do not need more employees or machines , due to robotic workers assistance.

I shall attempt to explain whether robotic workers will bring what positive or negative influence to our future working environment as below:

Advantages to robotic bring to working environment

What advantages thar robotic will bring to working environment? They may include: Many people fear that robots or full automation may someday take their jobs, but this is simply not the case. Robots bring more advantages than disadvantages to the workplace. They enrich a company's ability to succeed while improving the lives of real, human employees who are still needed to keep operations running smoothly. If you're thinking about investing in some robots, share the advantages with your employees. You might be surprised at how many of them are quick to support the idea.

1. Safety

Safety is the most obvious advantage of utilizing robotics. Heavy machinery, machinery that runs at hot temperature, and sharp objects can easily injure a human being. By delegating dangerous tasks to a robot, you're more likely to look at a repair bill than a serious medical bill or a lawsuit. Employees who work dangerous jobs will be thankful that robots can remove some of the risks.

2. Speed

Robots don't get distracted or need to take breaks. They don't request vacation time or ask to leave an hour early. A robot will never feel stressed out and start running slower. They also don't need to be invited to employee meetings or training session. Robots can work all the time, and this speeds up production. They keep your employees from having to overwork themselves to meet high pressure deadlines or seemingly impossible standards.

3. Consistency

Robots never need to divide their attention between a multitude of things. Their work is never contingent on the work of other people. They won't have unexpected emergencies, and they won't need to be relocated to complete a different time sensitive task. They're always there, and they're doing what they're supposed to do. Automation is typically far more reliable than human labor.

4. Perfection

Robots will always deliver quality. Since they're programmed for precise, repetitive motion, they're less likely to make mistakes. In some ways, robots are simultaneously an employee and a quality control system. A lack of quirks and preferences, combined with the eliminated possibility of human error, will create a predictably perfect product every time.

5. Happier Employees

Since robots are often assigned to perform tasks that people don't particularly enjoy, like menial work, repetitive motion, or dangerous jobs, your employees are more likely to be happy. They'll be focusing on more engaging work that's less likely to grind down their nerves. They might want to take advantage of additional educational opportunities, utilize your employee wellness program, or participate in an innovative workplace project. They'll be happy to let the robots do the work that leaves them feeling burned out.

6. Job Creation

Robots don't take jobs away. They merely change the jobs that exist. Robots need people for monitoring and supervision. The more robots we need, the more people we'll need to build those robots. By training your employees to work with robots, you're giving them a reason to stay motivated in their position with your company. They'll be there for the advancements and they'll have the unique opportunity to develop a new set of tech or engineering related skills.

7. Productivity

Robots can't do everything. Some jobs absolutely need to be completed by a human. If your human employees aren't caught up doing the things that could have easily be left for robots, they'll be available and productive. They can talk to customers, answer emails and social media comments, help with branding and marketing, and sell products. You'll be amazed at how much they can accomplish when the grunt work isn't weighing them down.

8. Cost reduce

The first and the foremost advantage of having robots in workplaces is their cost. Robots are much cheaper than humans and their cost is now decreasing. It's a fact that we cannot compare human abilities with robots but robotic capabilities are now growing quickly. For example, if you run an essay writing service, you can use robots to perform every kind of research related to any subject. Because robots are more active and don't get tired like humans, the collaboration between humans and robots is reducing absenteeism. The pace of human cannot increase hence robots are helping humans.

However,robots are more precise than humans; they don't tremble or shake as human hands. Robots have smaller and versatile moving parts which help them in performing tasks with more accuracy than humans. There is no doubt that robots are significantly stronger and faster than humans. Robots come in any shape and size, depending upon the need of the task. Robots can work anywhere in any environmental condition whether it is space, underwater, in extreme heat or wind etc. Robots can be used everywhere where human safety is a huge concern. Robots are programmed by a human; they cannot say no to anything and can be used for any dangerous and unwanted work where humans may deny to offer their services. For example, many robotic probes have been sent into space but have never returned. Robots in warfare are saving more lives and have now proven to be very successful. For example, in chemical factory environment, robots are now being used in the chemical industry and can, for example deal with chemical spills in a nuclear plant, which would otherwise pose a major health concern. Cost-effectiveness is one of the most sound arguments to be made for the case of industrial robots. Robots will reduce production costs by eliminating internal costs to compensate human salaries. Businesses are forecasting that their profitability will increase once they implement robots into production, or that they will have more financial mobility to invest in new products or technologies.

9. productive efficiency

Quality assurance is expected with the use of machinery in production. Industrial robots will be able to ensure consistency with mass production of manufactured products. The possible human error that assembly line workers pose the threat of will be removed. Optimized production efficiency means that a general manager will be able to have set quantity and quality standards that will be met by robots. Production quotas will not be jeopardized by low concentration, break time and employee injuries, among other things. The efficiency of production forecasts and supply levels will be increased with robots, able to be programmed to work at the optimal speed for a given plant. Limiting human work in hazardous environments, because manufacturing jobs often place workers at more physical risk compared to a lot of other industries. Lowering the level of a hazard presented to employees on the job is attractive to executives to preserve company reputation and minimize potential legal liabilities.

10. Reducing longer working hours

Typically people have to have breaks, get distracted and after time attention drops and pace slows. With a robot it can work 24/7, and keeps running at 100%. Typically if you replace one person on a key process in a production line with a robot the output increases by 40% in the same working hours just because a robot has more stamina and never stops. Robots also don't take holidays or have unexpected days off sick.

11. Increased profitability

By increasing the efficiency of your production process, reducing the resource and time needed to complete it, and also achieving higher quality products, industrial robots can thus be used to achieve higher profitability levels overall, with lower cost per product.

12. Improved working environment

Industrial robots are often used for performing tasks which are deemed as dangerous for humans, as well as being able to perform highly laborious and repetitive tasks. Overall, by using industrial robots you can improve the working conditions and safety in your factory or production process. Robots don't get tired and make dangerous mistakes, neither do they suffer from repetitive strain injury.Due to their high accuracy levels, robots can also be used to produce higher quality products which adhere to certain standards of quality, whilst also reducing the time needed for quality control.Industrial robots are able to complete certain tasks faster and better than people, as they are designed to perform these tasks with a higher accuracy level. This and the fact that they are used to automate processes which previously might have taken significantly more time and resources, means that you can often use industrial robots to increase the efficiency of your production line.

13. Improved Quality Assurance

Few workers enjoy doing repetitive tasks and after a certain period of time concentration levels will naturally decline. This lapse in concentration is known as vigilance decrement and can often lead to costly errors for the business and sometimes serious injury to the member of staff.Robotic automation eliminates these risks by accurately producing and checking items meet the required standard without fail. With more product going out the door manufactured to a higher standard, this creates a number of new business possibilities for companies to expand upon.

14. Increased Productivity

Using robotic automation to tackle repetitive tasks makes complete sense. Robots are designed to make repetitive movements. Humans, also by design, are not. The introduction of automation into your manufacturing process has many different productivity benefits, some of which are shown here.Giving staff members the opportunity to expand on their skills and work in other areas will create a better environment which the business as a whole will benefit from. With higher energy levels and more focus put into their work, the product can only improve, which will also lead to extremely satisfied clients.

15. Avoiding workers need to work In Hazardous Environments

Aside from potential injuries in the workplace, staff members in particular industries can be asked to work in unstable or dangerous environments. For example, if a high level of chemicals are present, robotic automation offers the ideal solution, as it will continue to work without harm. Production areas that require extremely high or low temperatures typically have a high turnover of staff due to the nature of the work. Automated robots can minimise material waste and remove the need for humans to put themselves at unnecessary risk.

Disadvantages to robotic bring to working environment

1. Increase unemployment rate and job loss

On working environment cost increasing aspect, where robots are increasing the efficiency in many businesses, they are also increasing the unemployment rate. Because of robots, human labour is no longer required in many factories and manufacturing plants. They can certainly handle their prescribed tasks, but they typically cannot handle unexpected situations.The ROI of your business may suffer if your operation relies on too many robots. They have higher expenses than humans, so at the end of the day you may not always achieve the desired ROI.

However, robots may have AI but they are certainly not as intelligent as humans. They can never improve their jobs outside the pre-defined programming because they simply cannot think for themselves. Robots installed in workplaces still require manual labour attached to them. Training those employees on how to work with the robots definitely has a cost attached to it.Moreover, robots have no sense of emotions or conscience. They lack empathy and this is one major disadvantage of having an emotionless workplace.Also, robots operate on the basis of information fed to them through a chip. If one thing goes wrong the entire company bears the loss. Where a robot saves times, on the other hand it can also result in a lag. It is, after all, a machine so you cannot expect too much from them. If a robot malfunctions, you need extra time to fix it, which would require reprogramming.If ultimately robots would do all the work, and the humans will just sit and monitor them, health hazards will increase rapidly. Obesity will be on top of the list. So there are advantages, but there are disadvantages as well. It is the twenty first century and we cannot work without machines.Humans are still considered far more efficient than robots when it comes to decision making powers, handling difficult situations, brainstorming, and generally bringing a sense of emotion and empathy into a workplace. Besides, you cannot rule out the significant role of humans in a business. After all, no machine can replace the human factor 'real employees' bring into a workplace. So, AI can raise unemployment and increase factory or shopping center or office working environment cost when their working environment are applied robotic to replace many workers, then machine electricity expense will also increase. Otherwise, human workers can not spend too much electricity expense in cost aspect.

Whilst industrial robots can prove highly effective and bring you a positive ROI, implementing them might require a fairly high capital cost. That's why, before making a decision we recommend considering both the investment needed and also the ROI you expect to achieve. Often the easiest way to get round this issue is to take out asset finance and the ROI of the robot more than pays for the interest on the asset finance.

This is typically the biggest obstacle that will decide whether or not a company will invest in robotic automation, or wait until a later stage. A comprehensive business case must be built when considering the implementation of this technology. The returns can be substantial and quite often occur within a short space of time. However, the cash flow must be sustainable in the meantime and the stability of the company is by no means worth the risk if the returns are only marginal. Yet, in most instances there will be a repayment schedule available, which makes it a lot easier to afford and control finances. Our downloadable automation payback calculator also has a finance scheme option so you can see how this would work for you.

On job loss increasing aspect, Job loss is by far the most significant opposition frequently brought against the use of robots in the manufacturing industry. Industry workers of all levels, from entry-level to veterans, worry about the security of their employment status, and the ability of their job to be replaced by a robot. This panic is more widespread in this industry compared to others because of the closer immanence of a robot takeover in manufacturing.

Macro effects are another topic that usually comes up with job loss. More "big picture" thinkers wonder how the national, and eventually global economy will be affected when manufacturing workers' jobs are displaced. How can this mass unemployment possibly be compensated for, and how can the robots' presumed success be limited from seeping into other industries. However, increased investment costs are a financial counterpoint to industrial robots, with the idea that manufacturing companies will rack up their debt investing in robotic technology. Firms that do not have the funding might even go bankrupt in an effort to keep up with industry trends rather than continue on with normalized operations.Hence, elimination of a whole labor class would presumably occur a bit of a ways down the road, but the implications of this point are too large not to consider. Bringing in robots to take unskilled labor jobs will place more pressure on the economy, education system, and financial market, just to name a few. The United States has always been associated with the grit and work ethic of its blue-collar workers, and robots are threatening to eliminate this aspect of the human population, with a take over of production jobs.

One of the biggest concerns surrounding the introduction of robotic automation is the impact of jobs for workers. If a robot can perform at a faster, more consistent rate, then the fear is that humans may not be needed at all. While these worries are understandable, they are not really accurate.The same was said during the early years of the industrial revolution, and as history has showed us, humans continued to play an essential role. Amazon are a great example of this. The employment rate has grown rapidly during a period where they have gone from using around 1,000 robots to over 45,000

2. Robotic can not perform better to compare human workers, when they need to work long time in any working environment

Robots need a supply of power, The people can lose jobs in factories, They need maintenance to keep them running, It costs a lot of money to make or buy robots, The software and the equipment that you need to use with the robot cost much money. Robots cost much money in maintenance & repair, The programs need to be updated to suit the changing requirements, the machines need to be made smarter, In case of breakdown, the cost of repair may be very high, The procedures to restore lost code or data may be time-consuming & costly.

Robots can store large amounts of data but the storage, access, retrieval is not as effective as the human brain, They can perform repetitive tasks for a long time but they do not get better with experience such as the humans do. Robots are not able to act any different from what they are programmed to do, With the heavy application of robots, the humans may become overly dependent on the machines, losing their mental capacities, If the control of robots goes in the wrong hands, Robots may cause the destruction. Robots are not intelligent or sentient, They can never improve the results of their jobs outside of their predefined programming, They do not think, They do not have emotions or conscience, This limits how the robots can help & interact with people. Robots can take the place of many humans in factories, So, the people have to find new jobs or be retrained, They can take the place of the humans in several situations, If the robots begin to replace the humans in every field, They will lead to unemployment.

Humans fear robots, Robots inspire two types of fear: firstly, that they might take over our jobs, and secondly, that they could take over the world, Robots will steal our jobs, Robots have the effect of increasing productivity rather than eliminating jobs.Robotics become increasingly present in our everyday life, with household robots, medical, industrial, on production lines, not to mention airports, banks, and hotels, So, Robots may dominate the human species. Robots can operate on the basis of information fed to them through a chip, when one thing goes wrong the entire company bears a loss.The robot can save times, but it can also result in a lag, It is a machine so you can't expect too much from them, If the robot has malfunctioned, you need extra time to fix it, which would require reprogramming, If robots would do all the work, and the humans will just sit and monitor them, health hazards will increase rapidly, Obesity will be on top of the list and less labour at workplaces.

3. Increasing training expense

Whilst industrial robots are excellent for performing many tasks, as with any other type of technology, they require more training and expertise to initially set up. The expertise of a good automation company with a support package will be very important. To minimise your reliance on automation companies you can train some of your engineers on how to program robots, but you will still need the assistance of experienced automation companies for the original integration of the robot.

In recent years the number of industrial robots and the applications they can be used for has increased significantly. However, there still are some limitations in terms of the type of tasks they can perform, which is why we suggest that an automation company looks at your requirement to assess the options first. Sometimes a bespoke automated system may give a better or faster result than a robot. Also, a robot does not have everything built into it, often the success or failure of an industrial robotic system depends on how well the surrounding systems are integrated e.g. grippers, vision systems, conveyor systems etc. Only use good trusted robot integrators to be sure of the optimum results if you do choose to use industrial robots.

IV

Can technology changes human social behavior

What does human network job mean ? Why may human network job be popular? Why human network job behavior may influence economy ?
Nowadays internet is popular to use. We can apply internet to find data , search any new things, even earn money. Why does internet may become huma network job source. For example, e-publish may be one kind of new human network job. Any authors may apply internet
channel to help them to sell electronic or paper books from e-publisher web store. They may apply facebook, you tub etc. any online
channel to promote themselves new books to let new readers to know whether when they may buy themselves favourable new topic books to read
from electronic publisher web store.

Thus, future electronic publisher industry may help any authors to build internet network platform to help them to sell and promote ot advertise their any one new electronic or paper book topic to let global any one reader to choose to buy their any new topic books from electronic publisher web store easily and conveniently. However, it implies that electronic network platform author may be one kind of future new human network job in our societies.

How electronic network platform author job may bring economy benefit in macro economy view? A person can have few friends, contacts and still be very influential if these few friends and contacts are themselves highly influential, e.g. one author must not need to know any one reader in global society. When they like to choose any electronic books from electronic internet network platform. They may become the author's any one topic book buyer, when they feel the author's any one topic book is fun and attract they make decision to buth the strange author whose the topic book from electronic book publisher's platform web store conventiently in short time. Although, they are strangers, they do not know themselves , but the reader can understand what it way that made Google from writing platofrm to create new creative mind and typing network job method to replace traditional hand writing book method for global authors. It will be one kind of new human network writing job.

Hence, global any one reader can apply an innovative search engine , such as google.com to find whether whom author personal new topic books are value to read from internet.
Then, the electroniuc publisher's web store may be new book store platform sale network to help the author to sell many electronic or paper books from electronic network platform in short time. So, internet may be future new network plaform to help global any one author to create network writing job absolutely. Furthermore, internet may be popular social media to help any one author to build goold relationship between his/her readers. It is one kind of new network, human network job. New authors do not need to buy many paper books to prepare to put in any one book shop warehouse. Their every book can print on demand to reduce out of book stock in any one book shop. They

may choose to sell either electronic books or paper books both from any one book publisher web store. So, electronic network platform may be one kind of good writing channel to help human authors to create income and it can also help authors to bring new creative mind and new topic fun content books to let readers to know and buy to read from electronic publisher network platform.

Why does human behavior may be one kind of new human network job to bring global economic advantages. ALthough, it may be free income or without inocme, but the person does the network behavior, his/her behavior may be bring advantages to influence many other people's health. For this case, when a worker in a coffee shop in an airport gets a vaccination aganinst the flu, it does not only helps him or her stay healthy, but also helps the many travellers who might otherwise have been inflected if that workers caught the flu. So, the externality , the result implies the vaccination of even a part of a community conveys benefits to the whole community. For example, governments pay special attention to the vaccinations of school children, teachers, health mothers, and the elderly, categories of people particularly susceptible not only to catching, but also to transmitting a disease.

It is not accidential that governments are heavily involved with vaccination . When there are externalities, free market, fail to persuade individual incentives with society's
their the worker's decision of whether to get a vaccine ends up attracting whether other people get sick. The workers might not fully take all these other people's potential suffering into account when making her or his vaccination decision.

As Stanford University does many suggestions, understand this and tries to help them make the right decisions and so providers free flu vaccines for its staff and students.

Small pockets of unvaccinated individuals can allow a disease to gain a spread more widely well-being. For example, parent weighing the costs and benefits of a vaccine for their child is not always thinking of the consequences of that vaccination to other people. THese are markets in which subsidizing or regulating behavior can make everyone better off. Because the reason for requiring that a child be vaccinated before enrolling in school is not just to protect that child, because each child's vaccination affects others via potential contagions.

Robots take our jobs behavioral and economy influences

Robot job behavior brings economy influences

If one day robots can replace human to do simple, even complex jobs. They will bring what influences to our global societial economy.The popular economic refrain declares that the global middle class is dying and robots will soon take our jobs, e.g. shopping center customer service jobs, library service jobs, cinema ticket sale jobs, restaurant kitchen cooker jobs, even, bus drivers, taxi drivers etc. public transport driving jobs, accountant, doctors etc. professional jobs. Whether it is beautiful or petty matter if our future societies have many human jobs can be replaced to do from robots. Businessman must may reduce to employ employees and reduce to pay salary or wage, when robots can be replaced to do their employees tasks. But, societies must bring unemployement rate rises , due to societies will have many people loss jobs when their employers choose to buy robots to serve their clients or do any office tasks or customer service or cleaning etc. tasks.

In micro economy view, employers may save money in long term, but in macro economy view, it will cause unemployment ratio rises , even crime rate rises when there are many people lose jobs in societies. These models of doom, though, fail to account for the hundreds of businesses riding the waves of change in their industries when robots may be invented to replace human to do many simple , even complex tasks in our future societies.

WE may image that one small factory needs to manufacture fishes canes to sell to supermarket, the small , cheaper stuff and higher margin parts of the fishes manufacture industry. Before, this factory needs to employe many human factory workers need to help every fresh customer makeing the perfect fishing gear,

designed for performance, durability, and cost in order to achieve to manufacture every fish cane in whole fished processing manufacturing stages. Every worker needs to spend about 15 to twenty minutes to finish every fish cane , till to delivery to any supermarket to sell. If this fish canes manufacturing factory can apply manufacturing robots to help them to finish any one working tasks , every robot can only spend five minutes to finish whole fresh fish cane manufacturing process. Thus, every robot can
help this factory save 10 to 15 minutes time to finsh every fish cane manufacturing process. IN fact, time is money, because when every robot can help this factory to reduce 10 to 15 minutes time to compare human worker. Then, this factory can finish about 20 fish canes in one hour if it can use robot to help it to manufacture fish canes. Otherwise, if this factory still use human workers to help it to manufacture fish canes, then it can finsh about 3 to 4 fish canes in one hour. SO, the manufacturing efficiency ensures that robots must help this fish manufacturing factory to raise fish canes number more than human workers. So, in robotic behavioral economy view, manufacturing robots must help this fish canes manufacturing factory to raise fish canes manufacturing number and deliver increasing number to supermarkets to prepare to sell every day. Robots can help this fish canes manufacturing factory bring manufacturing time saving, rising manufacturing efficiency, improving performance and reducing wages expenditure long time advantages in micro economy view. However, manufacturing robots can also bring disadvanages to society, e.g. increasing unemployment ratio, increasing crime rate,
this factory workers will lose jobs and income, they need earn social welfare from government and increasing government finance pressure in short time, even long time in macro economic view.

Stanford University graduate program in economics, Scott lecturer explained that "in demand and supply economic theory for robots supply and demand case, robots supply number increasing may influence human workers demand number decrease. It sometimes calls " the efficient frontier".

No specific human beings were mentioned in any of economics classes. As robots supply and demand in market case, They (robots) may be purely theoretical " agents" who reached to the most reasonable sale prices in order to persuade any one businessman buyer to make manufacturing robot buying decision whether robots can help him / her to bring how much saving time , saving money, saving cost, improving performance, efficiency economic benefit before he/she plans to reduce workers number when he/she decides to apply robots to replace human workers in his/her factory or office or any service department, e.g. cinema ticket sale service, shopping center customer service, shopping center cleaning , supermarket customer service etc. service or sale tasks. When robots can replace human to do any one of these tasks in any organizations. So, robots may be human worker agents who reached to prices the way robots would react to a software command. There was nothing that explained why some people thrived and others did n't or why truly brilliant, hardworking people could fail when much lazier folks succeeded." Having been admitted to the Stanford University graduate program in economics, Scott lecturer hoped to get his answers there.

How robots influence our future social changing? Using the right technology can be a boon to your business in this economy. For internet example, it is easier than ever to find well-matched customers all around the world, to stay in contact with them, and to more quickly design the products they want. If you focus solely on being cutting -edge, though you risk letting the technology take over what should be very robust relationships with your customers , employees, and colleagues. IN nowaddays society, technoligical advances and cutomation, personal relationships in business are more crucial than ever. I mean that robots can not replace human to serve clients to let them to feel more comfortable and passion more easily. For shoe shop case example, if the shoe shop apply one robot to serve its clients to replace human shoe salesperson to serve its shoe customers. Robots ensure that they can not persuade every shoe potential buyer to make shoe buying decision more easily when robots need to contact every shoe potential buyer. The reason is simple, because robots can not touch any one shoe buyer individual emotion very easier.

If the shoe buyer needs the robots to help him/her to choose any right shoe styles when he/she can not feel himself / herself can make the most right shoe style choice decision. The robots can not replace human shoe salesperson to make shoe style choice judgement more easily. They must need longer time to analyze whether which shoe style may be the most suitable to the shoe buyer. Otherwise, human shoe salesperson may attempt to make the most right shoe style choice decision to help any one shoe buyer to chooce the most right style shoe because he/she owns shoe style sale experience, shoe style knowledge, the most important reason is that they can feel every shoe customer individual emotion to touch whether he/she will feel comfortable or happy when they attempt to help every shoe customer to seek the most right shoe style in every shoe customer whole shoe searching processing. Othwerwise, serving robots are only one machine, they can not touch or feel every shoe customer individual emotion whether he/she feel comfortable or unhappy or happy when they need to contact them in whole shoe searching processing. Hence, I believe that some tasks robots can

not repalce human staff to do very easily. Otherwise, robots may bring disadvanatges to let any one businessman to loss his/her customers, due to robots can not touch every customer

emotion to compare human staff in service tasks more easily. Robots serving customer behaviors may cause money lose and customers number lose to the shop in micro economic view.

Intellectual human economic behaviors

What does intellectual human economic behaviors mean ? I believe that when we choose or decide to do intellectual behaviors, then our societies will be influenced to bring economic growth in consequence.I shall attempt to indicate pollution case to explain how and why eithet our intellectual or foolish behaviors may bring economic growth or recession in consequence as below:

On one hand, for air pollution social case aspect example, if we only consider to buy cars to drive for working aimr or holiday leisure aim. Then, our societies air will be polluted. Our health will be influenced to bad. Our car driving behaviors may cause global environment air pollution serously. In long tiem, global air pollution will bring our bodies health to be bad. Although, ourselves car driving behaviors may bring our driving travelling leisure enjoyment and comfortable feeling in short time, also we so not need to pay public transport fare often, but we need to compensate ourselves health economic intangible loss due to air pollution , when cars number increases, dirty air will cause ouselves health to become bad.

In the result, we will need to pay more medical expenditure when we are old age, due to ourselves bodies will become bad, due to we breathe global dirty air every day, due to ourselves cars pollute air in long time, e.g. 10 to 20 years, even 30 more without limited air pollution environment. So, driving cars behavior may be one kind of human foolish behavior and our foolish behavior may bring ourselves future long time medical expenditure absolutely.

One the other hand, water pollution social aspect, if we often keep much rubblish to pollute sea, oil exploration porcessing pollute ocean , ships gas pollute ocaen, then fishes will eat polluted food and drive dirty water, due to global ocean is polluted.

In fact, because human only to conside how to buy boats to carry on leisure enjoyment activities, or catch cruises to travel on the sea. Also, oil manufacturers only consider researching anywhere to find new oil exploration places to manufacture oil product, when their oil exploration processes pollute ocarn . Consequently, global fishes drink polluted warer or eat polluted food. They will have poison. SO, human will have high chance to eat poison polluted fishes, due to fishes are poison or are polluted.

So, human is doing foolish activities, we only hope to find oil exploration places to pollute ocean or we only spend money to buy ticket to catch ships to travel anywhere in global ocean. All of these human foolish behaviors will bring pollution to global ocean. On consequently, we will need to compensate to eat polluted or dirty or poision fishes, ourselves bodies health will be bad. In long time, we need have high chance to pay medical expenditure when we are old. So, pollution case may be one good example to explain how and why human foolish behavior may influence ourselves future need to compensate serious medical loss.

All of these human foolish behavior will bring pollution to global ocean. On consequently, we will need to compensate to eat polluted or dirty or poison fished , ourselves bodies health will be bad. In long time, we will have high chance to pay medical expenditure, when we are old. So, pollution case may be one good example to explain how and why human ourselves intellectual or foolish behaviors may influence future long time economic loss or economic growth or recession in micro and micro economic view.

On another water pollution aspect hand, if we often keep rubbish to sea, oil exploration processing pollutes ocean and ships' gas pollute ocean, then fishes will eat polluted food and drink dirty water, due to fishes will eat polluted food and drink dirty sea water because the global ocean is polluted seriously.

In fact, because human only consider how to buy boats to carry on any leisure water activities, or catches cruises to travel on the sea. Also, oil manufacturers only consider any where to find oil exploratin places to manufacture oil products from ocean, when their pol exploration processes can plooute ocean. Consequently, global fishes drink polluted water or eat direty food. They will have poison. So, human will have high chance to eat poison fishes.

Otherwise, such as pollutin case, it can infuence inflation or deflation. Consequently, the reason indicates supply and demand theory. If air pollution is serious, then we will consider health issue, global cars demand number may be influenced to reduce, when global cars number demand will reduce, global car prices and supply number will need to change to fall down in order to attract or persuade global car consumers choose to make car purchase decision.

Hence, global car manufacture number and car price will be

influenced to reduce, due to global air pollution issue. Consequently, deflation will occur because when the country citizen usually does not spend much extra saving money to buy car expensive goods. Money value will be low. Otherwise, if global cair pollution is not serious, human considers to buy cars to enjoy driving leisure lives. So, global car demand is influenced to increase , also global car price will also influenced to increase.

Consequently, gobal human will choose to buy cars to drive. Due to we accept to spend extra saving to buy expensive car goods. Car sale price and supply may be influenced to rise up. Money value is influenced to reduce. Inflation may be influenced, due to global car consumers number increases, we would not have extra money to spend easily. Car expensive goods expenditure influences our spending habit to avoid to make car purchase decision more easily. So, human intellectual or foolish activities may bring inflation or deflation consequency in possible indirectly in macro economic view.

On conclusion, above pollution case explain that how and why human intellectual or foolish economic behaviors may bring inflation or deflation consequency as wll as economic growth or recession consequency as well as any goods demand and supply increasing or decreasing consequency. It implies that human behavior may have indirect relationship to influence any goods demand and supply number to either increase or decrease result as well as any goods price will be influenced to increase or decrease in micro and macro economic view.

The relationship between social change and human behavior

Why does economic changes may influence human individual behavioral change? I shall attempt to indicate shopping behavior and staying at home behavior to explain their case and effect relationsip as below:

Human behavior can be influenced by economic change or economic change can be influenced by human behavior? Why does recession may influence consumers reduce shopping desire? In social recession suitation, it is possible that many people lose jobs suddenly, due to businessmen lose many customers. They need to make decision to reduce employees number in order to continue to keep businesses. Consequently, many firms (organizations) their employees may lose jobs. When they have much time, due to lose jobs, they will feel to avoid to spend too much time and money to go to shopping often. Many losing jobs people, they will often stay at homes.

So, they will reduce time to go to shopping, then non essential products won't their preferable choice purchase products. Hence, recession will change many losing jobs people their shopping or consumption desires to avoid to buy non essential products often . Usually when economic boom, many people have jobs to do because consumers number must increase when many people have jobs to do. Then, many people can accept to spend money to buy non essential products often. Many people feel spend time to go to shopping can satisfy their purchase of any kinds of new products useful psychology or desire. So, recession is one good example to explain it can influence many people do not like often to leave homes to go to shopping easily. Many people like to stay at homes, becaue they feel worry about spending too much shopping time when they leave homes. Their staying home time is one good negative shopping behavior example. So, economic change may influence human individual behavior changes , they have direct cause and efect relationship in behavioral economic view.

May human behavior influence economic change? Is it possible that human behavior may bring the country social economic change in macro economic or micro behavioral economic view ? I shall indicate publishing industry example. Do you feel that if there are many students feel learning is very important when they read many books or many of students feel interesting to read or they have reading new books in habit, then it is possible that the country will have many students like to spend time to go to any book shops to choose the books, they feel that they can help they learn new knowledge. Then the country will increase students number, they often spend time to visit any one book shop every week. Their visiting book shops behavior which may become their habits. So, the country will increase students number, they often spend time to visit book shops. Also, it implies that visiting book shops behaviors may be their behavioral habits.

So, when the country has many students often spend time to visit book shops , their visiting book shops behaviors may help any one book shop to raise books sale chance. So, the country's student individual often visiting book shop behaviors, their habitual visiting book shops behaviors must may assist help any one book shop to increase books sale number absolutely.

Consequently, any one book shop , its books sale bumber must be influenced to increase to increase because the country will have many students like or feel need visit book shops habit in order to choose any suitable books to buy to read at home in order to raise themselves learning effort. When the country has many bok shops often have many students visit their book shops, then their books sale number may be influenced to increase. It explain why student individual visiting book shop behavior may help any one book shop sale number increases also.

How human productive behavior may influence economic development

May any country which citizen behavior assist themselves country development? It is one cause and effect economic question. I mean that if the country itself citicen can not concentrate mind or energy to choose to do one kind of industry in order to let themselves country can bring the most benefit, then whether the counry itself economy can bring the most serious economic benefit. I shall attempt to indicate these countries themselves indistry choice to explain whether these countries themselves citizen productive behavior may help themselves countries to achieve the largest economic benefits. I shall indicate as below:

New Zealand farmer individual wine productive behavior

For New Zealand country example, this country concerns itself effort is foucs on farming agricultural aspect. So, this country has many farmers concentrate on farming agricultural aspect. May

New Zealanders choose to spend time to produce different kinds of wines, e.g. wine or red grape wine is for the people are eating meat, or they are eating dinner.

When these New Zealanders their behaviors choose to do farming or agriculture to grow and produce different kinds of taste of white or red grape wine drinking products job. Themselves grape agriculture behavior will influence these New Zealanders themselves, they can learn how to improve different kinds of grape wine drinking products in order to achieve every kinds of white or read grape wines taste improving aim during their white or red grape producing process.

Why can New Zealander every individual white or read grape wine producers improve their white or read grape wine taste more easily? In behavioral economic view, it can explain that why any one New Zealander white or read grape wine producer can be encouraged or excited or persuaded to concentrate nervous and energy and effort to learn how to improve their white or red grape wine products easily.

In fact, New Zealand is one agricultural food export country. It has good natural environment resource , e.g. land, seed to provide any one farmer to produce themselves any kinds of agricultrual food products, e.g. fruit, or wine food products. Because New Zealanders know themselves country has enough natural resource . So, in common, many New Zealanders choose to attempt to do farming agricultural jobs in order to export themselves any kinds of fruit or meat or wine products to overseas or sell to domestic in order to earn profit.

So, when these New Zealand farmers number has been increasing every year. This country farmers will feel themsleves competition between this New Zealand farmers themselves are serious due to they may feel New Zealanders choose to do agriculture businesses in order to export themselves different kinds of farming food to overseas or sell to local to earn profit.

Hence, when many New Zealand farmers feel that farmers number has been increasing every year. They will feel themselves competition is serious. They must need to spend much time and nervous and effort to research what method is the best how to produce the best taste of white or red grape wine products in order to let local or overseas wine buyers to choose to buy his/her producing white or read grpae products to drink.

Hence, in competition psychological view, may influence many New Zealand white or reaad wine producers had been beginning to change their learning behavior on researching what method is the best in order to produce the best quality of taste red or white wine products to sell in order to attract overseas or local white or read grape wine drinkers to choose to buy his/her wine products. Their behavior will focus on learning how to raising or improving white or read grape wine taste method more than only focus on producing a large number white or red grape wine products. They believe wine quality is more important to compare wine producing number. So, New Zealand wine producers themselves wine producers behaviors have been changing on concentrating on researching wine quality method aspect more then wine producing number aspect in behavioral economic view.

America high technological productive behavior

For America example, US is one high technological country, it owns many high technological knowledge talent inventors, e.g. computer science inventors. Hence, US must attract many diferent countries owning high technological computer inventors choose to go to US to develop their computer science profession career. Also, it seems that when many computer science inventors or professions choose to go to US to develop themselves computer science new career. In behavioral economic view, due to their leaving themselves countries choice, which may bring influence themselve country job behaviors need to be changed. They must need to adapt US new live. Because they will forgive their past computer science job. These computer science professionals need to spend time to adapt US new lives. They " past computer science job behaviors" will need to be changed to their new US any computer employer's new computer science job model.

Because their traditional computer science jobs needed to be forgot in their themselves countries. They will feel their old computer science job knowledge and behavior needed to change in order to let their US any one new of computer company employer feels satisfactory to accept their new working behavior in any one US computer organization.

So, on the other hand, many US computer company employer will feel that they must need time to accept any one new overseas computer science professions their working behaviors, their working attitude daily, because these foreign comouter science professional, their past computer working behaviors and working attitude must be different to US domestic computer science professions.

In behavioral economic view, these overseas computer science professions, their working behaviors and attitude must be needed to change in order to adapt any one US new computer company itself domestic or local computer science professional stafs themselves daily working behaviors and attitude because these overseas and local computer science professionals must need to team work together.

In behavioral economic view, it is only one way that foreign computer science professionals must need to change themselves past country traditiona daily working behaviors and attitude in order to cooperate with these US local computer science professionals in teams more easily.

Consequently, if these foreign compute science professionals can change their past working behaviors and attitude to let any one US local computer science professional feels to cooperate with them easily in short time. Then, the US computer company itself whole computer professional teams themselves efficiencies will be influenced to raised or improved by the changing past working attitude and working behaviors of these foreign computer science professionals. So, in behavioral economic view, only if US any one computer company hopes itself computer teams themselves efficiency can be raised or improved when it decides to employ foreign computer science professionals and US domestic computer science professionals. They need to work in teams together. They must need to let these foreign computer science professionals to know how to change their working behaviors and attitude to let their domestic computer science professionals feel easy to work together. Then, the US computer company itself whole team efficiency must be rasied or improved easily in short time.

- China share market investing behavior

For China share market example, economic development depends on financial market. Because if many Chinese have interest to invest to carry on shares buying and selling activities in orde to learn how to earn shares interest and share profit when the China shareholder can make decision to sell himself/herself shares in the the high price, then he/she can earn money when he/she can sell the China company's shares in the high sale share price position.

If China has many Chinese like to spend time to carry on investing shares activities. Themselves shares buying and selling behaviors will influence China has many companies can increase fund from many Chinese shareholders in order to have enough money to expand or develop themselves businesses in China in long term.

Consequently, when China can have many Chinese like to attempt to carry on buying and selling shares investing behaviors in China share market. Themselves buying and selling shares behaviors can help many Chinese companies have effort to increase enough money or capital in order to continue to do their businesses in long term absolutely. So, it explains why when many Chinese become shareholders , they can assist China will have many companies continue to develop their businesses if many Chinese like to carry on shares buying and selling investing behaviors in long time in China financial investment market nowadays in behavioral economic view.

Why has any individual country have many people invest share behavior which can influence the country's macro consumption desire?

I shall apply shares market buying and selling investment behavior to explaiin why shares investment behavior which may impact the country's overal consumption desire as below:

In behavioral economic view, I assume that when the coutry has many people have interest to attempt to carry on shares buying and selling investment behavior, then their frequent shares buying and selling behaviors which may bring negactive consumption desire or shopping desire of these shares investors their consumer behavior.

The reason is simple, when the country has many share buyers number suddenly been increasing rapidly. Consequently, these large group share investors must need to spend much time to research any kinds of company shares variations, whether when their share prices will rise up of fall down in order to achieve buying the company's shares in the lowest price and selling the company's shares in the highest price level in order to earn profit.

Basic on this reason, they must need to spend much extra time to research share prices changing behavior every day, e.g. one working person will wait to leave his/her job, after he/she can spend time to gather data to research the day's share price changing behavior after dinner. So, the working person's right time may be his/her share price market research behavior. Before he/she may spend his/her night time to go to shopping after dinner, but nowadays, he/she will fogive to do his/her shopping behavior before dinner or after dinner at hight sometime. He/she will make decision to spend much night time to turn on computer to click on share market website to research his/her share purchase choice to investigate whether his/her share price whether it rises up or falls down at the moment in order to make his/her share buying or selling decision at ever night time.

I mean the when the country has many people are share investors, their shares investment behavioral spenging time which will influence many shops lose customers at might often because the country will have many people feel need to spend night time to turn on computer or watch television to investigate share price variation. So, the country will have many people / share investors choose to stay at home in order to carry on share price variation investigation behavior, they need to listen share market update news from radios or watch the share market update news from computer or TV at home every night. Consequenly, they must reduce times to leave themselves homes at night. So, their shopping behavior also will be reduced. Because these share investors feel need to spend time to investigate share price variation news at homes which can bring economic benefits (high opportunity benefits) when they choose to forgive to leave homes to go to shopping times (opportunity cost) every night.

On conclusion, it seems that when the country has many people are share investors, then their share price investigating behavior may bring negative shopping emotion at night. Consequently, the country's any one shop may lose many customers from this share investor consumer group in behavioral economic view. Hence, when the country's share investors number had been increasing rapidly, it will influence any shops lose many customers from this share investing customer group at night frequenly in short time, even long time in behavioral economic view, because their shopping desires or shopping emotion will be brought negative feeling when they make decisions to spend much time to listen radios or watch TV or computers share price update nes at night. Hence, share market will bring negative impact to influence consumer shopping desire or negative shopping emotion in behavioral economic view.

Can technology influence human shopping behavioral change?

Nowadays, technological development has reached mature stage, whether technological mature stage may bring positive or negative shopping emotion influence to global consumers. I shall aplly internet inventin or ecommerce shopping channel tool to explain whether internet technology can bring postive or negative influence to global consumer behavior in behavioral economic view.

Internet is a good technological tool, it brings e-commerce business chance. In fact, commonly, global has have many businessmen choose to use internet channel to carry on their products transactions between global online-buyers and their electronic websites. So, global many shoppers had begun to feel online shopping is more convenient to compare visiting shops shopping. Their shopping behaviors have been changed from internet technological tool. Global has many shoppers choose to buy any products from any overseas or local businessmen their web stores. They only need to spend time to find any businessmen their webstores to choose the most suitable products to pay visa to buy from their webstores. at homes. So, in general, global had have may shoppers had changed their shopping behaviors from visiting shops to visiting webstores at homes often.

So, it seems that internet technological tool had influenced global many shops disappear, but internet webstores will be replaced their actual shops on streets. Some of businessmen either they choose webstores to replace shops or choose websotes and shops both or still keep shops only. Hence, internet tool influences global businessmen have three kinds of products sale channels to let globa local and overseas consumers to choose how to buy their products.

However, in fact, many of global shoppers, youngers and olders had begun to accept to buy any products from webstores. They feel to spend time to leave homes to visit shops , their shopping behaviors will be wasted time to not essential part to their daily lives. Hence, since internet technological invention, it had changed many consumers their traditional visiting shops shopping habit to change to buying products from webstores channel.

However, on the one hand, internet creates webstores ecommerce shopping channel to let global many consumers do not need to leave homes to go to shopping. It brings negative visiting shops shopping emotion to global general consumers nowadays. But on the other hand, it also brings positive visiting internet webstores shopping emotion to global general consumer nowadays. So, it seems that global many consumers feel that they often do not need to spend much time to go out shopping. Many global consumers feel convenient and enjoy to choose any products to buy from different internet webstores, when the online buyer chooses the most suitable product, he she only needs to pay visa card to buy the product from the online seller's webstore conveniently at home. Hence, online shopping can bring economic benefit to online buyers, e.g. avoiding walking time or spending transport fare to visit the shop to go to shopping, shortening or reducing shopping time to do another important matter.

On conclusion, global many consumers began feel online shopping can bring more economic benefits on shortening shopping time, avoiding transport fare spending aspect. So, online shopping will be popular shopping behavior for future long time. It may encourage global many shoppers can make rapid shopping decision in short time in order to carry on any products buying transaction to global any one online shopper in short time easily in behavioral economic view. So, global many businessmen had begun to build themselves one attraction webstore in order to persuade different countries consumers to choose to click themselves webstores from internet channel to buy any kinds of products in short time easily.

So, internet technology had changed consumers traditional shopping behaviors to build positive online shopping emotion as well as raise online sellers' any products sale chance easily in behavioral economic view.

Why and how human behavior may influence the country's economic growth or recession?

When one country has many people choose to do the same matter for one period, whether their behavior may influence the country's pvera; economic growth or recession . I shall attempt to indicate cases toexplain their relationship as below:

For flowing rubblish behavioral case example, do you feel that when the country has many people often flow rubblish on the streets, instead of their flowing rubblish behavior may bring streets dirty? But, their flowing rubblish behavior may explain that this country has people may have enough money to buy food to ear, or enough cloths to wear, enough bottles of water to drink, even they may have enough money to buy new television, radio, refrigeraters , washing machines, desktops or laptops electronic home products from old to new to use in order to satisfy their living needs. So, when they flow old electronic home products, their flowing old home electronic products behaviors may seem that they have enough money to buy other new home electronic products to replace old home electronic products to use at homes.

However, it seems thaat this country ought have many people have jobs to do. So, many of them, they can easy to make purchase decison to flow any old home electronic products and buy any new home electronic products to use . Because this country has many people have jobs to do. So, they can often not use old home electonic products to become rubblishs to flow on streets after they had bought any kinds of new home electronic homes.

In fact, it also implies that this country's economy grows rapidly. So, many businesses can glow up rapdly. When they expanded their businesses, they must need to increase employees number in order to let they help themselves to raise productivity or serve their clients absolutely. So, when the country has many businesses can grow up, it seems that its economy must be better or it is improved to compare past. Due to many different kinds of home electronic products had been often bought to use by this country people in this period. So, this country's any streets can be observed that expensive electronic home products were flowed on streets anywhere. then, this country will have many electronic home products sellers can sell their home electronic products very easily. When this country has many people can find any kinds of jobs to do easily. So, due to unemploymen rate had been decreasing.

In behavioral economic view, as this many electronic home products rubblish country case, we can observe this country may have many people have jobs to do. So, consumption number has been increased long time. So, cheap food, or expensive home electronic products may be rubblish on any streets. This country's people , their flowing rubblish behaviors may be explained that many of people have enough jobs to do, so they have ability to buy any good taste food to eat or buy any kinds of expensive electronic home products to use. So, this country's economy may be improved for this long period. So, in behavioral economic view, when this country can have many electronic home products rubblishs are flowed on anywherer in streets frequently. It seems that this country will have many people have jobs to do, so it causes they often change old home electronic products or replaced them easily, when they have enough income to spend to buy any kinds of new home electronic products to use at homes easily. Moreover, their flowing old electronic home products behaviors also indicate that this country has many people their salaries may be increased in possible from their emplyers. When this country can have many different kinds of home electornic products are sold. It means that this country's electronic home products needs or demand had been increasing, due to many people have jobs to do and income increases to excite their living of needs also improve. Consequently, this country may seem have better economic improvement. We can observe from this country's electronic home products rubblish increasing income in theis period.

On conclusion, this country ought experience economic growth at this period. So, " flowing expensive electronic home rubblish increasing number " may seem that this country's economic growth is rapidly in this period, due to many people have jobs to do as well as salaries increase in this period.

Technology how impacts human behavior changing?

Technology how influences human behavior to bring changing? For example, online share purchase and sale transaction from smart phone brings share investor can do share buying or selling transation in any where and any time conveniently, non manual driving auto vehicle, bring car owner feels comfortable and spends free time to do other matter, e.g. reading, listening mucis in himself or herself car freely. electrical energy vehicle can help car owner to reduce air polluton and it can brings the drivers do not feel drive long time in any journeys in order to avoid air pollution for environmental protection responsible car drivers in our societies. Thus, they will drive long time in any journeys when they can drive electronic energy cars to replace oil energy cars.

However, online technology can also bring consumers can choose to stay at homes to buy any things from seller individual online webstore conveniently. Such as online technology can bring shoppers do not need to spend much time to visit shops to buy any things. They can choose any kinds of products from any online sellers individual online webstores conveniently at homes. Online technology excite busy consumers can make purchase decision easily as well as it can help online sellers sell any kinds of products from internet easily.

In behavioral economic view, technology can change human behavior to be improved, it can let human feels comfortable, more free time ro use, rapid making any decisions, such as apply smart phones to make share purchase or sale transaction decision, online shopping decision, even travelling any where decision in short time, when the traveller finds the most cheap hotel accommodation room price and air ticket price frm any travel agent online tourism webstore, then the potential travel customer can follow the online hotel accommodation price and air ticket price data to make decision when to buy the air ticket from the airline travel agent or make decision when to prebook which hotel accommodation room to go to the country to travel from online travel agent tourism webstores. So, technology can encourage global any country travelers to make anywhere to trvel rapidly. If the traveler can find the country's general hotel rooms and airline tickets prices had been decreasing more sightly. The traveler may make travel decision to choose the country to travel in short time, then he/she can prebook the country;s any hotel room and airline ticket to pay by visa fraom the country's any hotel and airline travel agent webstores., before one week, even one month or more easily. Hence, online technology can also encourage traveler individual frequent travel times to be increased, due to global travelers can find any hotel rooms and airline tickets prices from internet conveniently at homes. They do not need to spend time to visit any airline travel agent to enquire travel choice country's hotel rooms prices and airline ticket prices. They can compare global travel of countries choices ' all hotels rooms and airline agents air tickets prices to make prebook airline seat and hotel room decision before one week, one month even six months early.

On conclusion, online technology can encourage global travelers can make travelling any where and when traveling time desicions easily. It can excite tourism industry develops in long time. Also, such as electricity cars invention can encourage environment protection car owners do car purchase decision easily, because they can choose to drive electronic energy cars to replace oil energy cars in order to avoid air pollution occurs easily. So, electronic cars can increase electronic car purchasrs number, due to many of environmental protection attitude of car owners can choose to drive electricity cars to bring air cleans, even non -manual driving cars can encourage lazy driving and free time driving car owners to choose to buy non-manual (artificial intelligent) cars to drive , because they can spend much free time to read, listen music or do any matters in themselves cars, they do not need to drive cars, robotic (AI) auto driving machine is such one non-manual driver to help them to drive themselves cars confidently. So, non-manual driving cars can attract lazy and enjoying free time driving car owners to choose to buy to replace traditional manual cars to drive easily. Moreover, online share transaction can help any share investors to make share buying and selling decision in short time easily. When they can apply smart phones technological tool to carry on share buying and selling activities easily. They can observe any share rising or falling price suitation from smart phones in any where any any time easily. So, smart phone technology can help global any shareholders to make share purchase and sale transaction easily. So, technology can encourage human makes decision in short time rapidly.

How and why employees behaviors may influence economy development?

In behavioral economy view,I believe the country's any organizational employees behavior may bring indirect relationship to influence the country's long term economic development. I shall indicate past manufacture industry social development period to explain their relationship. For many countries' past business activities had belonged to manufacturing industry, such as US, UK past before 1980 year, it focused on steel manufacturing and steel manufacturing related machine products. So, US, Uk developed countries manufacturing industries may be past main country's economic income sources. I assume US , UK past had one million number different kinds of industries. They ought had about seven houndred thousand number organizational businesses were belonged to manufactured industry. They may include:

Steel manufacturing and steel related machine manufacturing, e.g. vehicle manufacturing, home appliances, e.g. washing machine, television, radio, refrigerate cooler, heater, air condition etc. different kinds of different kinds of steel -related manufacturing machine, they were manufactured from US, UK steel machine manufacturers. So, US, Uk the other three hundred thousand number industry may be general service industry, e.g. hotel service, restaurent, cinema, public transport service, tourism lesiure , wine bar, supermarket etc. different kinds of non-manufacturing industries business organizations were operated in UK, US past before 1980 year.

So, in UK, US developed countries industry development history, they ought have high percentage of businesses belonged to steel related manufacturing machine and steel products. Also, in the past before 1980 year, US, Uk business employers , they employed many workers are manufacturing workers. They needed to spend long time to work in factories. They were skillful workers, and they are trained to manufacturing cars, washing machine, television, heater, etc. even steel itself different kinds of steel related products to prepare to deliver to their shops to sell to US, Uk local or overseas clients.

So, I believe that past UK, US ought employ many employees, they belonged to skillful manufacturing workers, manufacture increasing steel machine or steel related machine number of products rapidly daily. So, if UK, US had had many of these manufacturing factories owned high skillful workers, then their manufacturing steel-related machine or steel both kinds of products number must be influenced to raise rapidly. Consequently, their steel machine manufacturing products would been exported to overseas or would been sold to local both markets , they may be influenced to raise sale number. They (these manufacturing workers) needed to be trained to know how to manufactur these different kinds of machine products in the efficient teams and they ought to be trained to raise their efficiencies in order to shorten time to manufacturing many kinds of steel related manufacturing machine or steel itself products rapidly. So , if their efficiencies and manufacturing performance was improved, these US, UK any one manufacturing worker and their teams ought achieve raising productivities significantly.

Hence, when past UK, US manufacturing industry development period, if these two countries' any manufacturing factories could have many manufacturing workers could be trained to be skillful and proficient manufacturing workers. Then, in past every day to these factories workers, they ought help their steel or steel related manufacturing employers to raise any kinds of machine or steel products number in every team. So, when past in the manufacturing industry development, US, UK could have many factories' manufacturing workers themselves steel or steel related machine products manufacturing skill could be trained to to improve to any kinds of these machine or steel manufacuring products quality as well as their products number could be influenced to raise by themselves skillful improvement significantly every day.

Then, what would be influenced to occur to past UK, US manufacturing industry period? In behavioral economic view, when these two manufacturing industry developed countries, such as UK, US , if they had many factories workers can be trained to improve their skill in order to achieve any kinds of steel or steel-related machine products quality could be improved as well as products manufacturing number could be also increased absolutely. In consequence, past UK and US both countries ought increase themselves any kinds of steel and steel related machine products number to be supplied to themselves local shops to let local clients to choose any one kind of machine manufacturing products to buy easily as well as they could also export to supply overseas any countries to buy their different kinds of steel or steel related machine products to let overseas steel or steel related manufacturing machine product buyers, they can have many of these different kinds of these steel or steel-related different kinds of manufacturing machine from UK and UK these both countries easily to compare other countries.

On conclusion, I believe that past US, and UK macro manufacturing industry income GDP would increase significantly. So, they would have good economic growth performance because when many of these manufacturing workers themselves manufacturing effort could be improved. So, it explained when employees manufacturing abilities can influence economic growth indirectly.

Robots invention whether they can help organizations to raise efficiencies or inefficiencies?

In behavioral economic view, in any organizations, when the organization hopes its worker teams can raise efficiencies , the organization may choose to increase more workers number and/or it can provide training to improve these workets themselves skills in order to raise their efficiencies. For one warehouse example, when the warehouse increases many goods , they are needed to delivered these goods from the shelves to the delivering destination locations. If this warehouse supervisors feel these workers themselves goods delivery speeds are slow, which is possible due to this warehouse's workers number is not enough. So, this warehouse supervisor ought increase workers number in order to increase their goods delivery speed in order to deliver goods from the shelves to every indicated goods delivery destination in order to let any one lorry driver can transport the right kinds of goods and ensure the accurate goods number to transport to any one client home rapidly.

However, if this warehouse supervisor planed to buy several warehouse goods delivery robots to assist these warehouse workers to find the right kinds of goods from shelves and then deliver to the right destination location in the warehouse. So, these warehouse orkers can concentrate on counting the accurate goods number and ensuring the right kinds of goods in order to prepare to let lorry drivers to transport these goods to these goods of buyers themselvers homes rapidly. Consequently, in the first step, robots can concentrate on finding th right goods from shelves and delivers them to the right goods transportation of location destination. Then, in the second step, these warehouse workers can concentrate on counting the accurate goods number and ensuring the right kinds of goods in order to prepare to put them to the lorry. Consequently, when warehouse robots and warehouse workers can cooperate to work together, the most important, robots, can deal on finding the right kinds of goods and deal on delivering the accurate number of goods of job duty as well as these warehouse workers can only concentrte on counting the right kinds of goods number in order to avoid it has none any mistake of wrong kinds of goods and inaccurate goods of delivery number to be transported to the lorry and to deliver to any one buyer's home.

So, it seems that warehouse robots ought help any one warehouse worker to raise himself efficiency and avoid goods delivery of mistake occurrence easily as well as their help to warehouse workers that can let any one goods buyer feels their goods can be delivered to their homes rapidly. Moreover, warehouse robots can also help these warehouse workers to raise efficiencies because warehouse robots can help them to shorten goods delivery time between any one shelf and any one goods delivery destination of location in the warehuse because robots may help them to find the right kinds of goods from the right shelf in the short time. So, any one worker does not need to spend long time to seek anywhere is the right shelf location for the kind of goods when the kind of goods are needed to deliver to the buyer's home from lorry. Warehouse robots can help them to do this aspect of " finding the goods from the right shelf in short time job duty". So, any one warehouse worker only needed tospend less time to do the counting of any right kind of goods number and ensuring the right kind of goods job duty. Consequently, this warehouse 's any one worker, his any one kind of goods delivery time may be reduced, because robots' assistance and they may have more confidence to avoid mistake to deliver the wrong number of goods and/or the wrong kind of goods to any one goods buyer's home.

On conclusion, it seems that warehouse robots ought may help any one warehouse worker to raise efficiency for any one team in the warehouse as well as the warehouse any one supervisor does not need to spend much time to observe any one worker individual performance for " goods delivery job duty aspect" because their goods delivery job duty that had been replaced to do by these several warehouse robots. Robots can achieve the more accurate of right kinds of goods and the right number of goods delviery job performance to compare any one of human warehouse worker themselves right kinds of goods of delivery and right number of goods of delivery job performance. So, when robots can participate to cooperate with this warehouse's any one worker to do their goods of delivery job duty in this warehouse every day. Then, robots can raies any one of supervisor individual confidence in order to let they do not need to spend time to observe any one of worker individual whose goods of delivery job performane. They can concentrate on supervising any one worker whose goods transport to lorry in the final step in order to avoid to deliver wrong goods number and / or wrong kind of goods to any one goods buyer's home every day. Consequently, this warehouse's overall teams of their delviery of goods performance many be improved by robotss' participatin to goods of delivery task as well as this warehouse's oveall teams themselves efficiencies may be influenced to raise by robots' goods of delivery task participation.

www.ingramcontent.com/pod-product-compliance
Ingram Content Group UK Ltd.
Pitfield, Milton Keynes, MK11 3LW, UK
UKHW041630190726
13854UKWH00006B/2399

9 798888 330760